D0758798

MANAGING RISKS IN DESIGN & CONSTRUCTION PROJECTS

Ronald Saporita, P.E.

Library of Congress Cataloging-in-Publication Data

Saporita, Ron.
Managing risks in design & construction projects / Ron Saporita.
p. cm.
Includes bibliographical references and index.
ISBN 0-7918-0243-4 (alk. paper)
1. Building—Risk management. 2. Building—Safety measures. 3. Design, Industrial. I. Title.

TH153.S33 2006
624.068'1—dc22 2006003899

Table of Contents

Preface vii

Chapter 1: INTRODUCTION 1

1.1. Project Entities 3

1.2. Delivery Methods 7

1.3. Elements of Risks and Potential for Disputes 9

Chapter 2: IDENTIFICATION AND ANALYSIS OF RISK 13

2.1. Risk Management Process Overview 15

2.2. Project Triangles 20

2.3. Contracting/Delivery Methods 22

2.4. Scope Definition 27

2.5. Project Cost or Budget 30

2.6. Project Schedule 33

2.7. Resource Allocation 36

Chapter 3: RISK CONTROL 39

3.1. Overview 41

3.2. Cost Control 43

3.3. Schedule Control 46

3.4. Safety and Quality Assurance 51
3.5. Insurance and Bonding 52
3.6. Risk Management Log 54
3.7. Conclusion 55

Chapter 4: PROJECT "RIGHT START" 57
4.1. Overview 59
4.2. Partnering 59
4.3. Responsibility Matrix 60
4.4. Project Procedures Manual 61
4.5. Team Effectiveness 62
4.6. Project Health Assessment 67
4.7. Oversight 67

Chapter 5: RESOLVING PROJECT DISPUTES AND CLAIMS 71
5.1. Overview 73
5.2. Contract Provisions 73
5.3. Causes of Troubled Projects 76
5.4. Costs Associated with Delays 80
5.5. Claims Resolution 84

BIBLIOGRAPHY 87

ABOUT THE AUTHOR 89

GLOSSARY 91

APPENDICES 105

INDEX 155

List of Appendices

1. Goals Matrix
2. Project Risk Management Process
3. Record Keeping: Important Construction Documents
4. Types of Contracts
5. Cost Impact of Controllable Risks
6. Project Work Flow
7. Cost Tracking/Forecast Summary
8. Cost Trend Log
9. Construction and Commissioning Phase Risk Management Log
10. Partnering Practices Comparison
11. Dr. Deming's 14 Points
12. Project Oversight Checklist
13. Risk Memos
14. Oversight Process Diagram
15. Manshul Delay Damages Example
16. Acceleration Model: Productivity Issues
17. Measured Mile Productivity Damages Example

Preface

Capital investment projects, which include design and construction activities, pose risks to the owner, designer and contractor in achieving project goals such as safety, quality, cost and timeliness. Project management is the art of applying the necessary cost, scheduling and other variance identification tools to safely achieve a quality product, on time and within budget. The objective of design and construction project risk management is to ensure timely application of project control procedures for achieving all project goals, and to reduce the potential for claims resulting from disputes during or after construction.

Often the word "risk" is used interchangeably with "uncertainty." While it is uncertainty that must be managed, risk is a result of uncertainty.

No particular type of contract isolates participants from risks. An evaluation of risks involved in various contract formats, i.e., lump sum, cost plus, etc., and various delivery methods such as design/build and design/bid/build, is presented. Although contracts are not a comfortable subject for nonattorneys, knowledge of the risks from these is important for all project participants.

The Joint Engineering Society's body of knowledge for engineering managers (please refer to ASME's website for Engineering Management Certification International [EMCI]) includes project management principles and techniques as a subset of the "body of knowledge."[1] ASME's Professional Practices Curriculum (PPC), is intended to introduce young engineers to the principals of project management.[2] This text is intended to provide an understanding of a risk management "process" to support both EMCI and PPC. It contains a different perspective on managing for success than that in conventional project management texts; that of focusing on managing risks. *Risk management* is the application of various tools and techniques for identification, analysis, and control of risks. Each of these three steps is part of a continuing process throughout the project life cycle (concept through commissioning). The text is organized to emphasize the importance of fully evaluating the myriad of risks and their impact on project goals. Quantifying these impacts is essential for prioritization before applying various risk mitigation techniques. These techniques may be in the form of risk sharing, risk transfer, or risk attenuation.

The significance of project "phases," as well as the linkage between contingency and budgets, schedule and schedule allowance for each project phase, the need to plan for risk early in the project life cycle as well as continuous updating of the risk management plan is emphasized. Development of a comprehensive risk management process provides the basis for focus on key issues at the start and the inevitable changes to the baseline (scope, schedule and budget) plan as the project progresses.

The application of this project control technique encourages leadership qualities that seek to "raise the bar," not merely to attain it. In short, application of this technique to identify early, quantify accurately, and effectively attenuate project design and construction risks is a basis for continuous improvement.

Introduction

1.1 PROJECT ENTITIES

Management of risk is a common focus for many business enterprises. Risk is often considered in issues that concern cost, timing, safety or loss prevention, system reliability, overall performance, and product liability. Our focus is the utilization of various techniques to achieve the desired performance in design and construction projects, and mitigation of the potential for disputes or claims after completion of the project. Therefore, effective management of risk results in improved chance for business success, customer satisfaction, and claims avoidance.

Whether the application is for product development, manufacturing, or construction of facilities and infrastructures, the risk management concepts are similar. Because these projects involve the facilities and processes for virtually all technologies, including the associated buildings, structures and infrastructure (roads, rails, utilities, power plants, etc.), the focus is on the design and construction process.

There are generally three parties involved with the design and construction of a project: the owner, the design professional (referred to by the "lead" entity—which may be an architect or engineer) and the constructor (contractor). In many cases, these functions may be combined or further divided. For example, a project owner may hire one entity to perform both the design and construction, or the owner may decide to hire multiple designers and contractors to provide the design and construction services. For product development, however, the owner has the design and or manufacturing resources but may need to hire a contractor or contractors to construct the facilities.

Because of the diverse nature of construction in the United States, there are tens of thousands of owners initiating projects every year, and a larger number of design professionals and contractors. The cost of these projects range from the thousands of dollars to billions of dollars. There are a number of design and construction organizations that enjoy sales in excess of a billion dollars a year.

Many large engineering and construction companies, in the last 10 years, did not effectively manage engineering and construction risks and are no longer independent entities. As projects get larger and/or when the organizational relationships are new or complex, risks to success increase. *Construction Executive*[3] lists the most common causes for contractor business failure, where cash flow issues relate to budgets and the original schedule for deployment of resources. These are:

1. Growing too fast (cash flow and resource planning)
2. Inadequate capitalization (related to cash flow planning on all projects)
3. Poor estimating and job costing (related to cash flow planning on each project)
4. Poor cash flow (for all projects)
5. High employee turnover (resource planning)
6. Poor accounting systems (cash flow or resource planning for all projects)
7. Working in a new geographic region (resource planning)
8. Dramatic increase in single-job size (cash flow and resource planning)
9. Obtaining new kinds of work (leveraging customer satisfaction)
10. Poor purchasing decisions (cash flow and resource-supplier/subcontractor issues)

There are many notable project failures. Some have resulted in completion being delayed for years, and some have experienced cost overruns of 100% or more, introducing large disputes and claims. Outstanding construction claims in the United States exceed billions of dollars in a given year and generally take several years to resolve. Thus, the economic incentive to manage risks is obvious.

In contrast, those companies that manage risk well continue to appear or improve their ranking in the annual *Engineering News-Record* listings of top design professionals and contractors.

A brief discussion of entities and contractual relationships follows. Each entity has a primary responsibility and therefore an associated risk. That is, each entity has a "customer." The owner, for example, considers his customer requirements before investing. In this case, the customer is the end user of the product or facility. During design and construction it is not uncommon for requirements to change. Where possible, involvement of all entities in understanding the owner's customer's needs is an essential first step in understanding the project basis or scope. If, for example, the owner of a restaurant intended to lease the facility to another, the owner must have a clear understanding of the customer's requirements and may need participation of the customer in the design and construction activities to further develop the scope.

1.1.1 The Project Owner

The project owner is an individual or an organization that has a need to have a project designed and constructed in order to carry out their goals or obligations. It might be as simple as an individual wanting a new restaurant built in a period of 9 to 12 months, or as complex as the federal government wanting to build a nuclear waste processing facility, with expansion phased for a period of 20 years. As such, the owner identifies the scope of the project, in detail or with performance requirements, and the financial and timing restraints.

In some instances, the owner may have the resources for complete design. For larger projects, however, the project owner usually needs to procure the services of other entities to design, as well as to build the project. The project owner wants the project scope completed safely, on schedule, within budget, and of the required quality.

Thus, a key issue is that the owner understands, or will develop, the customer requirements to convey complete and unambiguous direction to the design professional or contractor. Of course, it is also understood that the owner will support the scope by providing funding and access to the work area in a timely manner.

1.1.2 The Design Professional

The design professional, alternately referred to as the designer, engineer, architect, or architect/engineer (A/E), is an individual or entity that is able to assist the project owner to meet its needs by designing (developing and specifying the scope, cost and timing) the project. In addition to the actual design services (preparation of plans and specifications), the engineer might also assist the owner in determining its full needs by participating in activities such as procuring specialty items, providing construction bid documents and postdesign completion services, and assisting the owner in starting up and operating the facility. The designer is expected to deliver comprehensive and unambiguous design or bid documents that include quality assurance requirements.

1.1.3 The Contractor

The contractor varies in capability, depending on the complexity of the project. The contractor may be as small as a one-man carpentry service performing work in a local community or as large as an international organization employing tens of thousands of workers across the globe. The contractor will construct the project, defined by the design documents, for an agreed upon fee basis or fixed price, and duration utilizing their own means and methods.

The following are some of the basic types of contractor organizations:

- *General contractor*—This is an individual or organization that typically contracts with the project owner to construct the facility. A party other than the general contractor typically performs the design of the project. The general contractor generally performs the work with a combination of direct hiring of individual craft workers and the use of subcontractors.

  Often, the work is bid as a fixed price and the general contractor is responsible to safely "deliver" the designed scope at the agreed to price, within the time allotted. This is referred to as being "at risk" because the contractor assumes the financial burden that may result from project delays or cost overruns (or both).

  In some instances, a reimbursable contracting method may be is used when the design scope is not sufficiently detailed or site conditions are uncertain.
- *Subcontractor*—This is an entity that typically contracts with a general contractor to perform a specified part of the work. The subcontractor may directly hire craft personnel to perform the work or use a subsubcontractor.

- *Specialty contractor*—This is an entity that provides specialized construction services usually involving a regulatory license or technical specialty [i.e., removal of contaminated soils, lead or asbestos abatement, x-ray inspection, electrical, plumbing, mechanical—or heating, ventilation and air conditioning, (HVAC), etc.]. They may be contracted directly to the project owner (and, as such are referred to as a "prime" contractor) or as a subcontractor to a general contractor.

1.2 DELIVERY METHODS

The successful completion of a project is referred to as its *delivery*. The "delivery method" describes the structure of the work relationship among the owner, designer and contractor(s). The method defines their roles and responsibilities as well as the risks involved. Typical delivery methods are Design/Build and Design/Bid/Build.

It is essential that the owner, engineer and constructor combine in a form (delivery method) that allows each to manage the risks for individual and project success.

1.2.1 Additional Participants

In addition to the general contractor and prime contractors, other participants in the delivery of a project include:

- *Construction manager*—When a project owner uses a number of separate, or "prime" contractors to construct a facility rather than hire a general contractor, and does not have sufficient resources to manage the work, the project owner may hire a construction manager (CM) to assist in obtaining bids, manage the construction and interface with the A/E. The owner's expectation is for the CM to provide comprehensive construction support services.

  Often, the construction contracts are between the owner and primes. The CM entity is not "at risk" to the owner (with which they are contractually bound) for cost, schedule or other issues. In some situations CM services can be provided "at risk," in a similar manner to the general contractor organization discussed previously. Here the primes contract directly with the CM. The contract form between owner and CM that is common for this situation is a *guaranteed maximum price* (GMP) (to be discussed later).

- *Owner's representative*—This term is typically used for an entity that provides a less comprehensive range of support services as compared to a CM. The services are not "at risk."

1.2.2 Design/Bid/Build

Here multiple entities are engaged to design and construct the project. In this method, the owner and designer may work cooperatively to ensure that all design requirements are finalized and individual bid packages are comprehensive to minimize scope development or omissions after bid.

1.2.3 Design/Build

This is an entity that has the capability to both design and construct the facility that the project owner requires, and be capable of turning over a facility ready to be operated. A Design/Build contractor can be referred to as a *turnkey* contractor or an Engineer/Procure/Construct (EPC) contractor. Often this type of approach is sought in order to accelerate completion (fast track) of a project with which the owner has experience. In this method, scope is often presented in the terms of facility performance standards, such that further design detail is necessary prior to construction. Here both design and construction risks may be assumed by this entity.

For any delivery method either reimbursable or fixed price contracts are utilized, depending upon the adequacy of scope definition. As scope becomes defined, the contract may be converted from reimbursable to fixed price or guaranteed maximum price type.

A variation of the above delivery methods is a design/build/operate contract where the contractor also operates the facility for the owner. This introduces additional risks, not addressed in this text.

1.3 ELEMENTS OF RISKS AND POTENTIAL FOR DISPUTES

All project participants assume financial and other risks in each design and construction project. Beginning with the design, or baseline plan, the following typical items must be considered:

- Customer buy-in
- Site access
- Safety requirements review (failure analysis, operational, construction, third party, etc.)
- Environmental impact (during construction and initial operations)
- Permitting
- Technology or supplier changes
- Resources—adequate contingency (time and money)
- Resources—labor (all aspects including management and trades)
- Clarity of drawings and specifications (single/multiprime)
- Quality assurance plans/testing and certifications
- Schedule activities of multiple prime contractors and subcontracting needs
- Communications with relevant stakeholders
- Validation of economic benefit to the owner

A thorough understanding of the risks involved in each delivery method and the variation in planned versus actual performance, is also essential. Even for similar projects in different locations, for example, the probability that individual design and construction tasks and costs will deviate from the expected (or planned) outcome must be considered. As such, project management is the art of securing the overall goals in the face of all the risks and changes encountered along the way. Success depends largely on carrying out the constituent tasks in a proven manner in a timely sequence, while deploying resources to the best advantage to maintain flexibility of approach.

Contracts, schedules, budgets, quality resources and safety issues must be well defined and understood by all entities at the beginning of their work. Trust and good communication depends on this. How many times have you heard the following?

- "I don't need to know about contracts, that's the lawyer's job!"
- "Schedules aren't followed, so why have one?"
- "We can fix it as we go along." (See figure.)

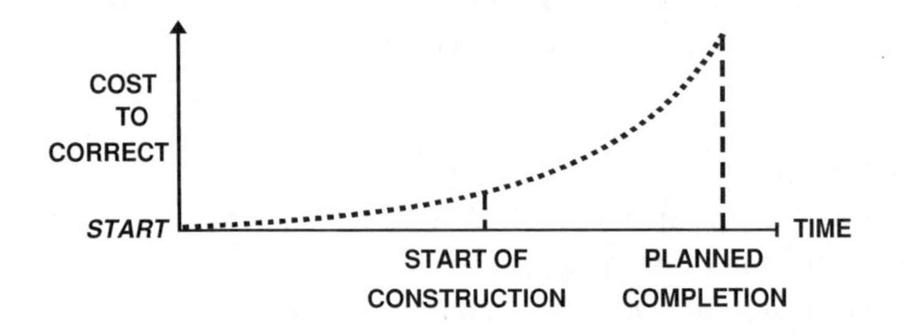

- "I can complete any job in a short period of time as long as I have the resources." (See figure.)

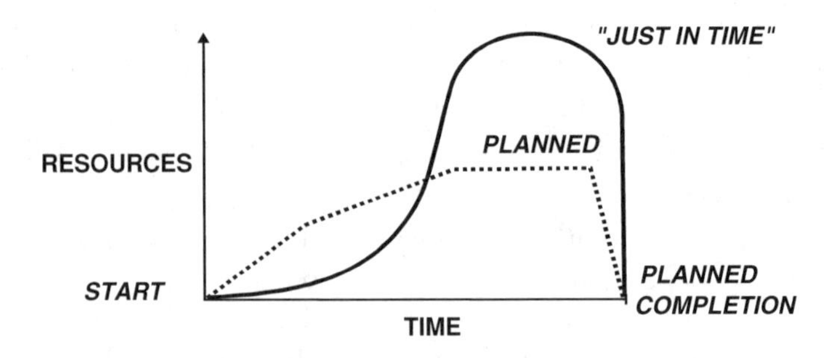

The roots of these sayings lie in a shallow understanding of how the goals of scope,* schedule, costs, quality and safety are integrated and contractually agreed to. Shallow knowledge is the path to failure.

All project goals must be quantified (have a numerical value). Goals may be elemental (apply only to cost, schedule, etc.) or composite (apply to all). Appendix 1 is a goals matrix that shows composite goals. The target for each elemental goal such as: satisfaction of design requirements, the budget in X dollars, duration in Y days; safety, in terms of Z lost time accidents/100,000 labor hours; and quality, in terms of % rework costs,

* *Scope* is defined as the equipment, materials and labor required in the design basis (for a designer) or the design documents and the means and methods required to complete the work (for a contractor).

are linked to show the desired outcome. Two examples are shown, one to achieve all goals (customer satisfaction), and one to exceed that outcome (continuous improvement).

Before the management of risks can be discussed, it is necessary to define risk. Often, risk is thought of as only the chance (or probability) that an undesirable outcome will occur. In fact, risk is measured in terms of the opportunity or loss being evaluated (such as $ or days), not merely the probability.

Specifically, risk = probability of occurrence × value, or amount, of the associated impact (consequence). For example, a 10% expected accuracy for a $10,000,000 budget could result in a consequence of loss of $1,000,000 in achieving the cost goal. If the probability of the consequence, based on experience, was 50%, the risk of overrunning the cost goal would be 0.5 × $1,000,000 = $500,000. Please note that while an accuracy is both plus and minus, denoting an opportunity to save 10%, in this example, the potential savings it is often not realistic (because of unknowns).

While some risks are addressed in a qualitative manner (e.g., the risk of budget overruns for incomplete design will decrease as the design becomes more complete), quantifying risks provides the most useful form of comparison. As can be seen, qualitative risks do not fit with the definition of risk. However, in some situations, because they are less time consuming or costly to develop, they are useful to identify risks for further analysis.

As stated, the overall goal of a successful project is to safely achieve the desired scope (and quality) with on-time performance, and to be within budget, it is necessary to address risks in a quantitative manner for as many foreseeable issues in design and construction as possible. This will allow for comparison with goals, prioritization of actions and evaluation of the results of improvement methods.

Because risks apply to the owner, designer and contractor, conflicts can develop as issues, such as scope changes, availability of resources, delays, etc., are identified and quantified. These may result in disputes or claims between the parties due to excessive costs to complete (by the contractor) or lost opportunity for the completed project (for the owner). In addition, there could be product (operational or environmental safety) or patent liability issues that result in claims. Finally, bonding and insurance claims could result in future premium increases or an inability to obtain certain types of bonding or insurance.

Thus, continuous management of risks is essential. The guiding management principal is that problems (or negative variances to a plan) should be "flagged" and addressed as early as possible. The only thing worse than bad news, is late bad news where something could have been done to mitigate an impact.

Identification and Analysis of Risk

2.1 RISK MANAGEMENT PROCESS OVERVIEW

Risk management contains the elements of identification, analysis and control. The focus of risk management must be on communication (and, therefore, understanding) by each party of the issues and risks involved. Each party must begin the project with an understanding and respect for each other's objectives and risks such that a risk management "process" can be implemented. Tools must be in place to evaluate completion progress against the initial, or *baseline* plan, to the satisfaction of each party. While this process is described in the context of design and construction projects (capital investment as opposed to maintenance projects), the application to product development and systems or manufacturing design can be developed by the reader.

In addition to individual (direct) risks to the three variable goals of scope, cost and schedule, it is important to recognize how the interrelationships between all goals introduce additional or indirect risks. For example, a change in scope for safety related issues (e.g., because students are occupying part of a building intended to be vacated) may introduce a risk to the cost or schedule goals. The relationships of key goals is discussed in Section 2.2.

At the start of a project certain risks are easily identifiable. Those that are not acceptable, are to be avoided. For example, if underground conditions could pose a risk to cost or duration, a detailed geotech report may be in order prior to starting design. Similarly, introduction of a new product element, without adequate testing would be avoided. Beginning with a design concept, identification of risk begins with evaluation of variances from the baseline plan, at the smallest definable task. For example, tasks such as the work of an engineer include the labor means and methods to develop a drawing, specification, etc. These tasks can introduce risks to timely completion by changes to the scope of work or to the planned design sequence.

Analysis of risk requires a quantification and prioritization of the issues. In order to prioritize issues, they must be compared on a common basis.

Since most risks impact the cost of a project, the interrelationships must be established. For example, a schedule or direct risk may be converted into the cost impact or indirect risk and compared with other cost risks.

Risk control includes transfer, sharing and attenuation of risk. Risk transfer is typically addressed by various contractual methods (performance bonds, etc.) and insurance (professional and product liability, builders risk, etc.). It is intended to address risks beyond the control of each of the project participants, or those risks not willing to be accepted by either party. The intention should not be to avoid all risks by attempting to transfer them to another party.

Risk sharing, typically defined in the contract, requires a "reasonable" assumption of opportunities for success. Success, for an owner, is obtaining the project or product within all goals specified. Success for a design professional or contractor is to make the anticipated profit, participate in ever-challenging projects, and satisfy the owner. There have been many instances where an owner unreasonably attempts to shift risk, thinking the risk is transferred (as opposed to shared) to the design professional or contractor (i.e., taking responsibility for all undisclosed site conditions yet not paying for the necessary site surveys to identify existing conditions). The results have been unanticipated costs for both parties. Risk sharing, therefore, includes both the reasonable contractual risks accepted by the entities (fixed price, for example), as well as the application of "partnering" (to be discussed later) principles to focus on the mutual benefit to all parties that accrue by reducing variances to shared project risks (costs due to changes or unknown conditions, for example).

Risk attenuation requires an understanding of contractual requirements to rapidly respond to project scope or other changes, continuous monitoring of cost and schedule variances and trends, implementation of effective safety and quality assurance plans, as well as flexibility in the application of resources.

The application of a continuous risk management process that incorporates identification, analysis and control should be considered at each phase of the project. This process is essentially a continuous improvement, or quality assurance process. Appendix 2 shows the project risk management process schematically for three project phases. Please note that the broader risk control term is shown as component of the planning phase process when risk transfer and sharing issues are typically being evaluated, while the attenuation term is shown as a component for the design and construction phases, subsequent to establishing a contract, or contracts.

With early application of this process to positive and negative impacts to budget procurement issues, progress schedule of critical tasks, scope changes, changes in sequence of work, quality and safety, etc., risks can be managed. Positive variances are important also because it may be possible to exceed the goals or provide offsetting allowances as the project progresses. The focus is not only on issues but also on the timeliness of the decision-making process. Because projects are typically broken into phases issues must be resolved as soon as possible within a phase, and prior to moving to the succeeding phase, thus the dotted line "loop" shown in Appendix 2.

This chapter explores the issues relating to identification and analysis of risks. Examples of these steps are included in Chapter 3 with the discussion of risk control.

2.1.1 Risk Identification

Identification of project risk begins with an understanding of the business plan (private sector) or program plan (public sector); especially the elements of scope, cost and schedule—even prior to entering into a contract.

Given that these project goals are variable, it is first necessary to identify the risks in each. At each phase of a project negative variances in budgets and schedules are the most common form of identification of areas to be investigated. The impact of one variable on other variables also needs to be identified so that further analysis and risk attenuation activities can be developed. Time-tested project control tools (such as scheduling software, cost forecasting reports and documentation procedures) must be integrated at each phase of the project. A risk management log, discussed later, can be used to supplement these typical project control tools.

Identification occurs by a deviation from a *baseline* or based on historical information. These deviations either have obvious consequences (will increase cost or duration), or may not be able to be quantified without further analysis. Not only is the nature of the risk important, but also the responsibility. For example, site conditions (nature of soil, unknown buried components, etc.) that differ from the defined basis must be addressed in terms of who will pay for any impact should mitigation fail. Thus, it must be established (in the contract) that payments will be consistent with the differing condition, if the contractor is not responsible for creating or discovering the condition.

Since identification of risk is the starting point in the process, and essential for notification and communication with the project participants, the importance of documentation should be noted. Without a clear definition and presentation of predicted and evolving project risks, resolution will not be possible. Should an issue be unresolved, and a claim develop, effective documentation will be essential for resolution of the claim—even years after the project is completed. In addition to the written contract and baseline schedule and budget, other important documents are given in Appendix 3, Record Keeping: Important Construction Documents.

2.1.2 Risk Analysis

Similar to identification of risks by use of project control tools, analysis involves quantification of the impact on the project goals. If the impact is identified in qualitative terms, it must be further developed in quantitative terms. Rating, or prioritization, is often used to help focus further risk analysis and attenuation activities. Figure 1 is an illustration of a qualitative risk rating system. Example categories that are analyzed are:

a. Potential scope change impacting time and direct costs, such as undisclosed conditions
b. Potential scope change impacting direct costs, such as drawing errors
c. Design developments/clarifications with potential scope and time implications, such as at the time for subcontractor shop drawing review
d. Changes in code requirements during the project

Because of the greater impact on goals, the lowest Roman Numeral (I) would be the highest priority to investigate further.

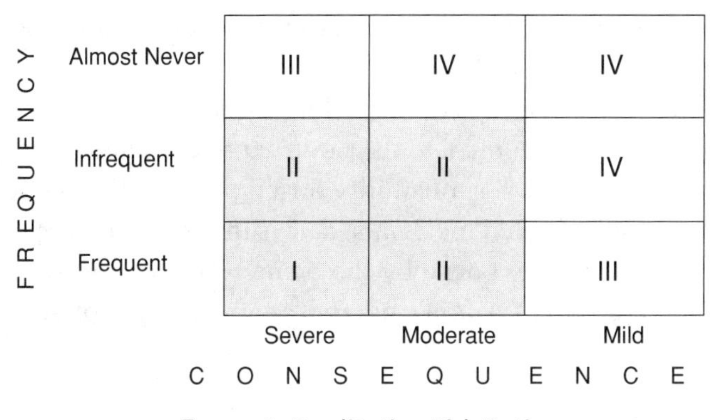

FIGURE 1 Qualitative Risk Rating.

As stated, only quantitative values can be related to the business plan performance goals of project cost and schedule duration. While quantitative values cannot be applied with 100% accuracy, the significance lies in the probability of the variation. For projects based on existing technology, historical information is typically considered sufficient for estimates and durations. Costs and durations, therefore, can be considered discrete variables that can be quantified with a reasonable degree of certainty. The application of a contingency, to address this lack of certainty, will be discussed later.

For new technology or development projects, where experience in technology or means and methods may be lacking, individual task costs and durations may be considered to be *random* variables. A Monte Carlo simulation or other probabilistic approach is often utilized. Monte Carlo simulations, for example, use three values and assign a distribution relationship to each, for each cost item or task duration. Variation and uncertainty as well as the dependencies of one task on another must be evaluated. With this information a random variable analysis of the work would be developed to quantify risk. The output of this analysis is a relationship between risk acceptance and budget contingency or schedule total float. Further understanding of probabilistic methods can be obtained from the chapter on risk simulation[4] and other publications. It is clear that the uncertainty for these projects results in risk acceptability beyond that acceptable for most projects. While the risk management process for such projects is similar, the following discussions will focus only on projects for which elements of scope experience exist. The significance of this discussion is that risks for new technology (even in one piece of equipment) or new means and methods must be carefully analyzed.

In summary, analysis of risks is performed by the use of contract documents and/or project control tools. Contract documents might describe methods to address the responsibility for changes to the scope of work, quality or cost of preferred materials, delays in regulatory approvals, etc. These risks are typically borne by the owner. Project control tools typically provide a means for analysis and quantification of schedule and cost related risks.

2.1.3 Risk Attenuation

Once the variances from project goals are quantified and responsibilities assigned, informed team decisions may be made and the necessary management support provided to mitigate the impact. Time is of the essence in engineering and construction, and, therefore, for the decision making process. The importance of quick and early action should be obvious. Changes to scope made at the start of a project phase, can generally be accommodated with the least cost and time impact.

In general, for the three project goals of scope, cost and schedule, a change in one will impact at least one of the others. Therefore, it is important to prioritize the goals at the start of a project. Effective control of scope changes, adequate budget contingency, and baseline schedule *total float* (time allowed for a task to slip without impacting the completion date) are essential in reducing the variances of the three goals.

Examples of priorities of goals are: time is critical and costs are of secondary concern (within reason) or budget is critical and completion time is of secondary concern. If cost and time are of equal priority, this *balanced* option requires well-integrated project control tools to answer the question: If a project is late, and the owner must pay for acceleration to recover lost time, can extra costs be justified?

2.2 PROJECT TRIANGLES

For illustrative purposes, the three key project goals of scope, cost and timing can be visualized as the sides of a triangle. If each side is fixed at the start of a project, the impact of trends can be visualized. While a variance is a change to a goal that is committed or spent, a trend is a probable change to goal that is not fully committed. Trends are used for early identification.

Assuming adequate budget contingency and schedule total float, only one or two of the three project goals are typically maintained throughout a project. Because these goals are variables, they may be either dependent or independent (those with a higher priority). Often, scope is an independent variable with cost and time as dependent variables. As discussed in the previous section, there are other instances where time or cost can be the priority variable. The following examples illustrate the interrelationship of the goals. Please note this discussion assumes that safety and quality goals are constants, not variables.

- Increases in scope results in an increase in time or cost to complete, or both.

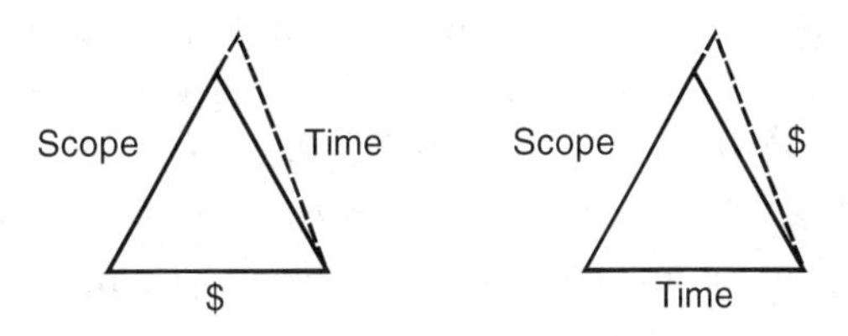

- Increase in time (project duration) usually results in an increase in cost to complete (for all entities) or requires a decrease in scope by the owner (via value engineering,* or a functionality reduction) to maintain costs. Usually a decrease in time does not provide a savings to the owner in construction costs.

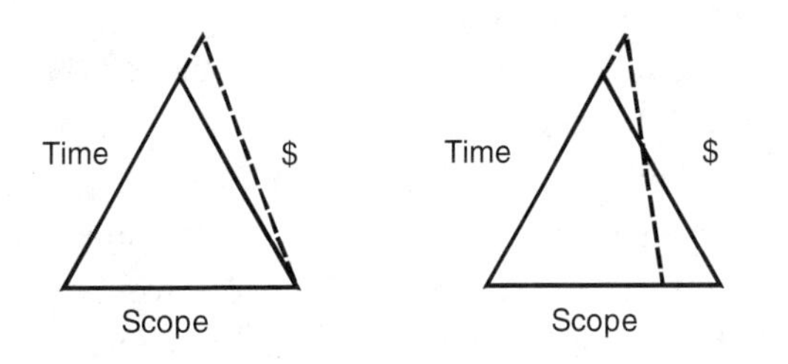

* See Section 3.1.

- Increase in costs beyond control of the contractor can require a decrease in scope (by the owner) to maintain the budgeted cost.

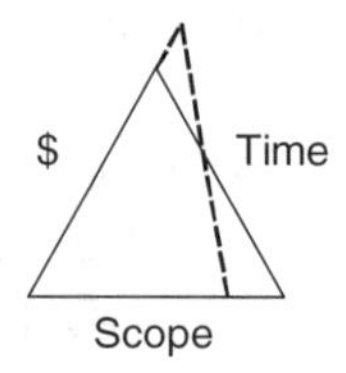

2.3 CONTRACTING/DELIVERY METHODS

An important decision in the project execution strategy is which delivery method should be utilized and how the type of contract will support the method. A misunderstanding of the basis for selecting a contract type or delivery method could result in an unnecessary increase in risk. For example, should there be multiple contracts for design and construction, (design/bid/build) or just one (design/build), and should the contracts be reimbursable (cost plus) or fixed price (lump sum)? The elements of these decisions are derived from the priority of goals (relative importance of cost or schedule) and how they are to be managed. While there is much attention to design/build, today, it should be noted that one approach cannot be used for all projects, the importance of prior experience of the design/build team should not be overlooked, and no contract isolates participants completely from risk.

The owner has a number of choices in delivery approach and contract types. Key elements affecting the choices include:

1. A focus on the business need (goals) relative to cost and schedule. For example, does the need for a short duration (fast track schedule) justify increased costs?
2. The risks of cost and/or schedule increase relative to completeness of scope definition, and how to manage them. This is sometimes dictated by *perceived* risk avoidance (design/build).
3. The turnover or commissioning requirements. How extensive will training be, or will build/operate requirements apply?
4. The view of the contracting method from owner's, designer's, and contractor's points of view. How high can we make the probability of a win-win outcome?

5. Where "management responsibility" will lie, and how interfaces and timeliness of the decision making process among the owner, design professional and contractor will function.
6. The construction industry climate and potential contractor interest.
7. Project funding source. Since this is the life-blood of the project, all participants should understand how a project is funded and the constraints and timing of approval for potential changes.

As mentioned, there are two basic contract types: fixed cost (or lump sum) and reimbursable cost (or cost plus) with many variations (guaranteed maximum price, unit prices, etc.) and combinations to these two basic types. Professional services are often contracted on a reimbursable basis or as a percent of the total project cost, while construction services are most often contracted on a fixed cost basis. The most common options include:

- *Cost plus*—This type of contract works best when the scope of the required work cannot be established at the time of entering into the contract. The project owner agrees to pay the designer or constructor their costs plus a fee. The negotiated fee could be a fixed amount or a percent of the actual costs.
- *Lump sum*—This type of contract works best when the scope of work (quantity of work and duration of project) are clearly stated in the contract documents and not subject to major change. This type of contract would be used for a building project when the design is complete and the project drawings and specifications can be included as part of the contract. It must be noted that a fixed price contract is not a zero risk (for the owner) option. Any contract is dependent on the completeness of the scope documents (errors and omissions) and possible unknown field conditions, site access restrictions or reasonableness of the duration.
- *Unit price*—This type of contract works best when the type of work is known, but the quantities cannot be determined prior to the work being performed. This contract form might be used for the installation of a sewer line when it has been determined (from soil borings) that there will be some rock and some earth in the area

where the sewer is to be installed. However, the quantities of rock and earth cannot be determined, with certainty, until the work is actually performed. With a unit price contract the bid documents usually contain the designer's estimate of the expected quantities of each bid item. Many contracts have a clause that allows for an adjustment of the unit price in the event the quantities vary by a stated amount from those estimated quantities shown on the bid documents. Schedule considerations may be included by incorporation of a schedule of planned durations.

- *Guaranteed maximum price (GMP)*—This type of contract works best when the design is based on conventional means and methods, but insufficient scope information is available for fixed price bidding at the time of award. Typically, the *guarantee* (price cap) contains a reimbursable contingency to cover the contractor's uncertainty. There may also be an agreement for the owner to share in the savings or overruns against the GMP. In either case, the contract must clearly spell out what costs are to be reimbursed and what costs are fixed.
- *Convertible contracts*—unlike GMP contracts, convertible contracts usually remain reimbursable until the scope is fully defined. At that point a fixed price arrangement may be implemented. This could allow the owner to establish its cost exposure with greater certainty and possibly reduce the reimbursable contingency available to the contractor.

Other considerations for selection of contract types include:

1. The designer's or contractor's experience with similar projects, in the same location, and the experience of personnel assigned.
2. Effective and equitable use of monetary incentives (bonuses) or remedies (liquidated damages, etc.).
3. Owner access and quality control requirements. Will there be access restrictions, quality checkpoints, inspections, etc., to incorporate in the schedule?

Appendix 4, Types of Contracts, is an overview of various contract forms and typical advantages and disadvantages. As can be seen, there are many variations of contracts. A construction attorney should be consulted before selection.

2.3.1 Qualitative Risk Comparisons

Often decisions on contracting/delivery methods are based on a qualitative basis. Table 1, is an excerpt from Appendix 5, Business Roundtable's Report A-7, Contractual Arrangements, Construction Industry Cost Effectiveness Report,[5] which summarizes the risk of typical variables in various contract types to owners. It is based on projects where the owner is reasonably certain of the scope required to satisfy his needs and the design is complete. The "cost impact to owner" is a qualitative assessment of the "risk" to the owner. While this assessment is not universally agreed to, it is presented to show an example of the application of qualitative risk analysis.

Since this assessment is qualitative, its usefulness is limited unless key indicators are quantified and an effective contracting strategy is developed. As discussed, the interrelationship of variables must also be considered. Because no contract shields either party from risk, owners, design professionals and contractors must agree to the quantification of risks. Considering the example of where schedule considerations and cost considerations often pose competing risks, if a delivery method requires construction bids without design completion (because of time constraints), one would expect a higher bid cost due to scope uncertainty. A method must be established to quantify this expectation and provide an adequate (owner's) budget contingency.

It should be noted that public sector owners often include a No-Damages-For-Delay (NDFD) clause in their contracts for risk transfer. Some contractors also include a NDFD clause in their subcontracts. Therefore,

TABLE 1. Cost Impact to Owner

	Fixed price	Reimbursable
1. Labor productivity	Medium	Medium
2. Scope	High	Low
3. Indirect costs	Low	High
4. Quality construction	Medium	Medium
5. Safety	Medium	Medium
6. Schedule	High	High
7. Labor relations	Low	Low
8. Project management	Low	Low

quantification of delay cost risks would be an important consideration for a contractor or subcontractor. Of course, such a clause has limitations and a time delay risk can affect both entities.

2.3.2 Forms of Agreement

For common projects, it is often cost effective to use standard forms of agreement. In the U.S., both the Engineers Joint Contract Documents Committee (of the American Consulting Engineers Council, the American Society of Civil Engineers, and the National Society of Professional Engineers) and the Architectural Institute of America offer standard contracts for use.

By using standard document language to define the respective responsibilities of the parties to a project, there is substantial advantage based on "test of time" experience of interpreting such language. But as noted on the cover sheet of each document, they have important legal consequences and should, before their employment, be reviewed by an attorney. It is important to remember, however, that the specific project attachments to these forms contain the details upon which the project requirements are defined. This is as important as the basic contract and associated general conditions language.

Overseas, the United Nations Commission on International Trade Law (UNCITRAL) and Fédération Internationale Des Ingénieurs-Conseils (FIDIC) provide resource documents for construction of industrial works.

Once the contract type and delivery method are selected, there is little opportunity to change. Risk transfer issues, such as insurance and bonding requirements, are often fixed with the selection. Risk sharing issues, such as change procedures and notification provisions remain important considerations throughout the project. Because failure of timely resolution of disputes can result in failure for either party to achieve their (and common) project goals, consideration is often given to incorporating a neutral third party into contractual dispute provisions for significant issues.

Therefore, the early opportunities for risk management lie in establishing effective risk sharing and transfer, ensuring completeness of scope definition, resource identification (specific project management experience and bidder qualifications, as examples), and clarity of schedule, cost, quality, and safety requirements. In other words, establishing a contract format that incorporates the complete scope, creating comprehensive cost and schedule baseline plans, and instituting effective project controls at the start of a project.

2.4 SCOPE DEFINITION

While contract and delivery method risks were previously addressed, the primary focus of uncertainty, and hence risk, lies in adequate scope definition. Because a typical project plan development consists of discrete phases, risks can be related to completeness of scope definition.

At the inception of a project, the end user (owner) may not be 100% certain of the scope of a project to fulfill his needs. The end user's desired *program of needs* may also continue to evolve during the design and construction phases. Similarly, environmental issues (i.e., asbestos abatement, lead abatement, etc.) may not be fully defined. The variability in design requirements, operational needs of the owner's business, or end user, and unknown conditions, therefore, play an important role in risk identification.

For new technology virtually all design considerations contain uncertainty which implies risks. Typical product or equipment design considerations include:

- Function
  - Weight
  - Size
  - Shape
  - Surface/style
- Kinematics
  - Motion
  - Loads
  - Controls
- Mechanics of Materials
  - Stress/strain
  - Deflection/stiffness
  - Fatigue

    - Wear
    - Corrosion
    - Noise
    - Thermal performance
- Economics/standards
    - Interchangeability/tolerances
    - Volume/recycle
    - Life cycle
    - Liability/safety
    - Reliability
    - Environmental
    - Customer documentation

These issues are often evaluated via various component tests.

Since systems are composed of many products or components, risks can be created from multiple sources.

Risk assessment techniques such as Hazards and Operability (HazOp) Technique, Failure Modes and Effects Analysis (FMEA), Fault Tree Analysis (FTA) and Event Tree Analysis (ETA) are utilized to establish criteria for reliable and safe designs of systems and components for new technology or systems, and to aid in establishing the scope of such projects.

For example, HazOp is a qualitative assessment technique used to determine an undesirable system performance while FMEA is a semiquantitative assessment of system performance. FTA is a quantitative assessment of various failure modes (causes) to determine the probability of an undesirable event (top event) of a system. The ETA is a complementary quantitative assessment used to analyze multiple outcomes, with different levels of reliability or safeguards, for an undesirable event. Please refer to the Safety and Risk Assessment Module[6] for additional information.

Scope that satisfies the owner's needs is developed into a project plan, which contains the variables of cost and time. Each of these variables contains a degree of risk, both individually and collectively. Risk identification should consider historical project data (for similar projects and location) as well as current external factors (regulatory requirements, market conditions, etc.). Depending upon the phase of a project, certain expectations may be anticipated. Appendix 6 depicts a project work flow diagram containing a description of a multiphase project. While this example is for a process-type industrial project, similar examples can be prepared for transportation, infrastructure, schools, hospitals and other types of projects.

It is the nature of scope definition and, therefore, project risk management, that prior to funding, a suitable number of phases be established by the owner for each project. The following description of project phases, shown in Appendix 6, is for projects with which prior experience exists (not "first of a kind"). In general, projects can have up to six phases. The left side of Appendix 6 describes the various phases and design cost expectation. The key activities listed on the right depicts the elements that contribute to improvement in definition, and, therefore, reduced cost and schedule risk as each phase is completed. These phases are discrete steps leading to decision points, or management interaction with the development process. They may be resourced (or contracted for) separately or combined, and are determined by experience with each type of project.

Phase 1 Planning ("feasibility") study—Typically used to provide a basis for a preliminary economic evaluation. This requires an understanding of the project (with reference to a duplicate or similar project) and may typically require expenditure of up to 5% of the total design cost forecast.

Phase 2 Conceptual ("assessment") design—Typically used by businesses to provide the basis for a formal position paper to support the business plan. This requires a complete process definition or program of requirements (to firm the scope) and may require expenditure of up to 10% of the total design cost forecast.

Phase 3 Basic design—A balance of overall project duration and cost risk to support funding approval. This requires some detailed design effort and may require expenditure of over 20% of the total design cost forecast. This is the point at which design/build projects may typically be bid.

Phase 4 Detailed design and construction planning—This phase results in completion of construction design/bid/build documents that includes identification of long delivery items, constructability and value engineering issues. This may require expenditure of up to 95% of the total design cost forecast and construction funding for long delivery items.

Phase 5 Construction, procurement and precommissioning—Requires the expenditure of the balance of design cost forecast to support completion.

Phase 6 Commissioning—Warranty transfer, start-up and operational requirements. A long-term operating and/or maintenance agreement may also be applicable (build/operate).

It is essential to note that the contract documents developed prior to construction award must include all requirements (such as inspections needed for certificates of occupancy, operating manuals, as-built drawings, training, etc.) for successful commissioning and transfer of operational responsibility. As such, the construction phase must *begin* with a plan to focus on these elements (as opposed to waiting to the end of a project to address these issues).

2.5 PROJECT COST OR BUDGET

As shown in Appendix 6, during the course of a project, from the planning phase through construction, more accurate estimates are developed to reduce the risks involved. Typically, information from informal quotations, or historical information, engineering judgment and experience factors are used by owners or designers in assembling cost data during early project phases. As more scope definition is developed, cost data are obtained from suppliers, manufacturers and contractors.

The estimate accuracies shown below are the results of the engineering activity of the project phase. These, or a variation, are used by some owners and designers to evaluate the cost accuracy, and resulting budget implications. As another example, the Association for the Advancement of Cost Engineers has issued Recommended Practices[7] for cost estimate classification that contains a slightly different system addressing the same concern.

The estimate for each phase is used as a baseline of the cost for the subsequent project phase. It is axiomatic that improving estimate accuracy, as each phase proceeds requires additional level of design detail. There is no free reduction of risk; it can only be masked (or worse, increased) by shifting the design responsibility from the designer to the contractor.

Historically, four classes of prebid estimates have been used for well-defined scopes, with proven (not first-of-a-kind) equipment, experience in a location, and available materials and experienced human resources. An example of a well-defined project would be one that was previously designed and constructed such that, this "duplicate" project benefited by incorporating identified changes in the design phase. An example of a project with evolving definition would be a first-of-a-kind size, complexity or technology.

Project Phase	Estimate Description	Purpose	Class	Accuracy Range
Planning study	Preliminary estimate	Feasibility analysis	I	±25–40%
Conceptual design	Factor estimate	Early stage assessment	II	±15–25%
Basic design	Budget estimate	Budget for funding approval	III	±10–15%
Detailed design and construction planning	Definitive estimate	Construction cost control budget	IV	±5–10%

As can be seen, the variation in accuracy is substantial for Class I and II estimates. Accuracy must be reflected in an allowance for design development; a specific sum included for each expected item of purchase, subgroup, or service. It allows for normally expected growth during the evolution of a design. These allowances are components of an owner's contingency and cover ongoing refinement of the project definition such as:

- Increases due to absence of testing for unknown conditions
- Increase in equipment and material costs due to final layout
- Increases in quantities as details in minor systems are finalized
- Increases due to specification clarifications, enhancements or additions

As such, this is the estimate of the work as described in the documents (you can't estimate what you can't define). This accuracy range does not address additional contingency components such as material or labor escalation (which may be transferred to a contractor in a fixed price contract), allowance for possible minor scope changes, or potential design errors or omissions.

Since 100% design completion is rarely attained when going out for bids, final completion costs above the bid cost should be expected—even for a lump sum bid. Therefore, an owner's project budget must include an adequate contingency to account for both the accuracy of an estimate and an allowance for unforeseen risks and changes.

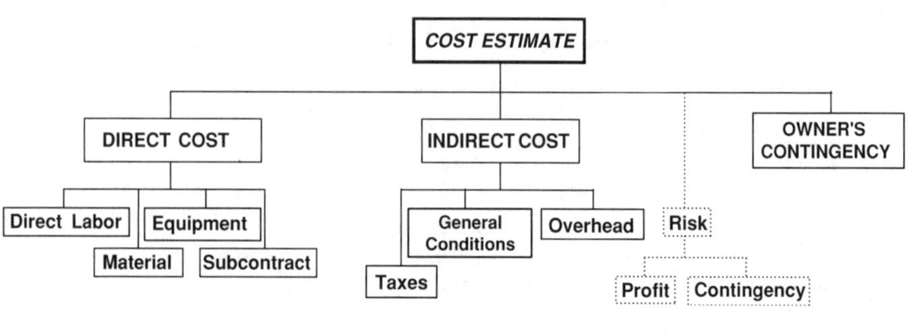

FIGURE 2

Figure 2 depicts the components of a budget. All should be self-explanatory, except General Conditions. General Conditions are costs related to performing the work in the field that do not result in a permanent feature of the final asset. These include trailer rental, small tools, supervision costs, telephone, etc. The dotted boxes shown, are contractors expectations unknown prior to bid.

After the detailed design is completed and the contract documents are finalized, the project owner is ready to go to the marketplace and find the true cost of constructing the project. While the designer has been estimating the costs throughout the design phase, until a contractor submits a price for the work, which may be backed up by a contract and a performance bond (discussed later), the owner can't be sure of the price of construction. The contractor's risk requirements cannot be estimated and will only be evident at time of bid. Because of this, the more the design is complete and clearly presented, the better the opportunity to obtain competitive bids. A well-coordinated (for the trades) set of design documents is essential for competitive bidding, and will provide the information for progress measurement during construction. Omissions added after contract award are more costly due to a loss of the competitive bidding opportunity.

It can be seen from the previous section that in order to fully support the contracting method, the appropriate level of design completeness must be obtained. A "fast track" method, typically involves the award of construction, as well as design completion, beyond Phase 3 (discussed in Section 2.4). As discussed above, timing (fast track or not) and budget cost accuracy often favor opposite approaches in degree of design completion prior to construction award, requiring a decision to be made on the relative importance of each (timing vs. cost).

As will be discussed in Chapter 3, all budgets must be formatted consistent with resources and associated tasks, such that measuring progress of the tasks can be performed.

2.6 PROJECT SCHEDULE

Any project without a schedule, understood by all, is doomed to failure. Thinking in terms of sequential phases ensures that all decisions are made and all resources are allocated for efficient movement into each succeeding phase. The initial approved (by the owner) project schedule represents a fully integrated (all phases) project with additional definition for the current project phase. It is the planned, or baseline, for future reference until changes are authorized.

The objectives of scheduling multiple tasks (from a Work Breakdown Structure, or WBS) include:

- Provide for owner input via milestones, approvals or other activities
- Identify calendar completion dates and constraints
- Determine cash flow requirements
- Determine resource requirements
- Refine estimating criteria (validating the budget)
- Obtain improved project control capability via what-if analyses capability

- Provide effective communication for buy-in by all participants (including subcontractors)
- Provide a basis for easy revision and comparison to the baseline schedule

Because the construction phase (with multiple tasks and resources) has the largest potential risk to the project goals, the construction schedule is a prime source of risk identification and analysis. It shows progress or lack thereof. All participants must contribute to, or *own*, the schedule and all revisions as the project progresses.

In order to own a schedule, the information must be provided in a user-friendly manner. The demands and the varying requirements for timely information by different participants often results in the need for different formats, or levels of detail taken from the schedule database for owners, subcontractors, etc. This is done to enhance communication so all parties can review issues, and time related risks can be identified and quantified. Of particular importance to all participants, is the focus on the project's critical path (or longest path of related tasks). Incorporation of cost and resources associated with activities, especially near or on the critical path, allows discussion and decisions regarding correction of negative schedule impacts. Whether from scope changes, manpower or material issues, these impacts could introduce a time related cost risk to the contractor or owner, or both.

Unit price tasks may also be included in fixed price and reimbursable contracts. Often, they represent a small part of the scope/cost of such work. However, it is essential that unit price tasks, that may potentially affect completion, be incorporated in the CPM schedule for the project. This will facilitate identification and quantification of unforeseen impacts.

In order for the project schedule to be a useful tool to identify, analyze and prioritize the related risks (and associated costs), certain factors must be incorporated:

1. The Critical Path Method (CPM) schedule should clearly demonstrate how the work is intended to be completed. The work sequence and time durations should be actual intended ones, not simply something to satisfy the contract requirements and use up all available time.

2. Total float should be available with flexibility built into sequencing to "recover" lost time. Total float is the schedule equivalent of budget contingency.
3. The schedule should be sufficiently detailed (WBS) to clearly show the use of major equipment (crane for steel erection), sequence of activities by task and location, etc., as well as regulatory milestones.
4. Major schedule activities should be cost and/or resource loaded to adequately evaluate options to sequencing and durations planned.
5. The actual progress of the job should be tracked and recorded in accordance with the original (baseline) schedule or authorized changes.
6. Near term (two-week) look-ahead bar charts (by trade) are used to supplement the CPM.

Tracking progress must be related to a "physical progress" measuring system and not merely the percent of resources expended from the budget or construction duration expended from the schedule. During the design as well as the construction phases, physical progress may be measured against individual tasks in the WBS. For example, during the design phase, design man-hours can be assigned to the number of expected design deliverables (drawings and specifications). The amount of documents completed, therefore, can be used to determine physical progress. Physical progress can then be compared to costs and time expended for the tasks. The application of physical progress measurement is an important part of cost and schedule control for all design and construction activities. Within reason, the greater the detail, the greater the accuracy.

In order to (timely) identify variances, the schedule should be updated periodically throughout the life of the project. Usually this is a once a month occurrence, with distribution to all involved parties. After each update, all parties should ask themselves the following three questions:

1. Where are we now?
2. Why are we here?
3. What should we do about it?

Running a what-if analysis to a cost or resource loaded schedule can determine the most effective path forward. The schedule update will show the planned resources utilized to achieve the actual completion. In many situations, the resources actually utilized to this point, may exceed those planned. Thus, an evaluation of productivity at this point provides the basis for forecasting the resources needed to complete the project within the time planned, or to accelerate the duration.

As a project nears completion, the focus of scheduling changes from construction completion by area, to completion by systems or phased operational needs. Pre-commissioning and construction completion must be treated as concurrent activities. As such the CPM, while essential for completion tracking, must be supplemented with a pre-commissioning or operational transfer schedule. Without this, time risks cannot adequately be identified nor analyzed.

In summary, schedule variances can be identified when completion progress is regularly compared with the approved baseline schedule, and supplemental schedules. Negative trends (increase in project duration) may require re-sequencing or additional resources to minimize the variance from the baseline, and must be evaluated and acted upon promptly.

2.7 RESOURCE ALLOCATION

In addition to labor included in the cost estimates, the significance of availability and cost-effectiveness of qualified supervision and support services cannot be underestimated. First, the owner should evaluate their staffing, or those selected to represent them. The owner must select the right person(s) or entity—at the right time to manage or oversee the job. This is perhaps the most critical step the owner will take. Most importantly, how will requests for additional funding be approved? Who will be granted the authority for changes to funding requirements or extensions of time?

An owner may be faced with a decision to contract for outside services or to manage a project with its own employees. A choice of organizational structure, such as an Owner's Representative or a Construction Manager may be considered. A lack of understanding of the relationships of these two types of management and their responsibility for interactions with the design professional and contractor could lead to unanticipated conflicts. Even with the use of qualified companies the owner may face a significant management risk. Inexperienced people, or those with poor negotiation skills (which are extremely important), may be offered to an owner when the market is in tight supply of such people. This management risk could impact all project goals. Budgeting, scheduling, quality and safety knowledge, and experience with comparable projects and management role, should not be overlooked.

As a project progresses from conceptual design through commissioning, the management method introduces risks during phase transition. For example, if a single project management team is responsible for a project from concept to commissioning, it is expected that there would be less scope and related cost risks as compared to when different project managers are assigned to each of the phases. (No single point responsibility and more opportunity for issues to "fall in the cracks" or be misinterpreted.) While this is difficult to quantify, an owner must consider this in the context of allocating funds for a comprehensive or well-qualified team.

Whether or not insurance or bonding is provided by subcontractors, owners often require approval of all subcontractors before they begin work on site. As part of the approval process, some owners perform an *integrity* risk audit for all construction entities. This audit consists of:

- *Legal screen*—confirmation of a clean record regarding labor law violations and satisfaction of litigation issues
- *Competency screen*—review of previous customer satisfaction, subcontractor "network", and projects of similar complexity
- *Financial screen*—review of Dun & Bradstreet (or similar) report, impact of pending litigation/arbitration, and operational/financial controls

Finally, in certain locations, engineering and craft resources are in short supply or union contracts may be scheduled for renewal. In either case, these risks should be identified in sufficient detail so alternate sources of labor may be considered or potential cost increases factored into project decisions.

Risk Control

3.1 OVERVIEW

Risk control contains the elements of risk sharing, risk transfer and risk attenuation. Those risks that cannot be controlled, are by default, accepted.

Risk sharing is based on an evaluation of opportunities for success for each party. For example, if the contract contains unit prices, each party must anticipate a "worst case" shift in the proportion of low-cost to high-cost quantities and the associated change in costs or time impacts for differing conditions. A quantity variation clause could provide a mechanism to share the risks.

It is common for entities to utilize insurance policies and surety bonds to provide risk transfer. The contract may also transfer some time or cost risks to the contractor via various clauses (No Damages For Delay) or the delivery method.

Although risk sharing and risk transfer are elements included at the time of entering into a contract, a thorough understanding is necessary to manage the impacts as the work progresses.

The third element of risk control, risk attenuation, is based on mitigating all risks including those that were "thought" to be transferred or shared. As discussed previously, attenuation actions require completion of the previous risk management process actions, identification and analysis.

Quantification of risk allows for prioritization of resources to apply control actions to project goals. The impacts of certain decisions, such as owner resource allocation, cannot be quantified nor controlled. Others, such as cost and schedule risks are related to various practices, or circumstances, such as incomplete scope definition, unknown site conditions, and unanticipated delays can be quantified. It is these risks that require mitigation or attenuation to the extent possible.

It was shown above that some project impacts may create risks to more than one goal. This discussion focuses on attenuation of risks that occur after a contract is awarded. Those related to market conditions (shortage of labor, etc.) are not included.

Timely identification and analysis is essential to attenuate risks and avoid disputes. Prior to contract award, scope issues need to be checked and rechecked, and a constructability analysis performed. During the design phase of the project, a review of the designer's documents for completeness and constructability allows adjustments to be made while the project is on paper. This early review ensures that contractors bid from complete drawings, specifications, and schedule/phasing elements to minimize potential for changes. Of equal importance is that it creates an interface with the end user to confirm acceptance or rejection of any changes and addresses other owner concerns.

Often the terms *constructability* and *value engineering* are incorrectly associated. For clarity, value engineering (VE) is a multi-discipline, systematic and proactive function targeted at the design itself. The objective in using value engineering is to develop a facility or component design that will yield the lowest life-cycle cost or provide the greatest value while meeting all functional, safety, quality, operability, maintainability, durability and other established performance criteria. Thus, the term value engineering is associated with a project under development.

As noted above, constructability decisions are focused on construction both prior to and post design completion. The intent is to save construction time and total cost without compromising other project objectives such as reliability, operability, maintainability, durability, and appearance. Consequently, constructability decisions are oriented toward:

- Reducing total construction time by creating conditions that maximize the potential for more concurrent (rather than sequential) construction activities of all subcontractors
- Reducing work-hour requirements by creating conditions that promote better productivity
- Reducing costs of construction equipment (and tools) by reducing requirements for special equipment and creating conditions that promote more efficient use of available equipment
- Creating the safest workplace possible, since safety and work efficiency go hand in hand

These are achieved by:

- Coordination of the contract drawings. All too often, one of the architectural, civil, structural, electrical, plumbing or mechanical drawings are modified, but the change is not carried throughout.
- Comprehensiveness check. Documents are reviewed to ascertain essential sections or details which have been left out of a bid package. In addition, multiple prime contracting requires a similar check to insure the information is comprehensive.
- Review of the specifications for errors, omissions, conflicts and coordination.
- Review of the special conditions of the site, and site logistics to inform the contractor(s) on the limitations and need for special construction requirements. This may include special requirements for working hours, milestone tie-in dates or site access or laydown areas.

After contract award, there are typically three types of documents that are used to identify and analyze risk. These documents are:

1. Contract with associated scope definition
2. Cost report/forecast
3. Cost loaded CPM baseline schedule and update forecasts

Since these documents focus on a baseline and associated variables, they do not address areas of uncertainty. A fourth document, a *risk management log*, which summarizes the quantification of risk on a comparative (cost) basis, will be introduced later. This is an important component of the risk management process.

3.2 COST CONTROL

Once a budget has been established, each element of expenditure must be compared with the budget and means applied to reduce or eliminate cost increases (value engineering, splitting or combining bid packages, etc.).

Cost variance analysis is an essential part of a continuous risk management process. Cost variances are identified as costs are committed (awards and payments). By comparing the actual cost and earned cost (based on physical completion) to the baseline budget, variances are identified. Earned costs are developed from a budget task grouping (i.e., installation of piping in an area) where each element of work is quantified (i.e., feet of various types and sizes of pipe to be installed) and the unit cost of each is identified. As a quantity of work is complete, the value, or earned value, of the work can be computed. This, then, allows a comparison with the actual costs incurred, which will permit productivity or other variance evaluations. Utilizing the individual variances, a forecast for the entire task group may be made. Applying this to all tasks in the work breakdown structure, weighted on a total budget cost basis, will yield an overall percent complete and provide the basis to develop a forecast to compete.

Without a reduction in scope, resulting from value engineering or some other technique, negative trends (increase above budgeted amount) can only be relatively mitigated by cost savings from another budgeted item, or utilization of a contingency (which is an allowance for risk or uncertainty). Appendix 7 is a contractor Cost Tracking/Forecast Summary for the construction and commissioning phases of a project. This summary shows variances compared to the budget and includes a physical completion evaluation so that productivity may be factored into the forecasting methodology. This is a simplified format. Typically a summary would also include columns that show changes from the prior period; information which is vital to timely identification of variances.

The amount of contingency remaining and variances are critical indicators. Forecasting not only includes evaluation of the variances discussed above, but also identification of trends. Trends are an early warning that a reasonably expected cost increase or decrease will occur. Depending on the magnitude or how long it takes to quantify, it may be added with the related line items (as a committed cost) in the variance analysis or incorporated as a trend via an adjustment to the Contingency, in Appendix 7.

A Cost Trend Log (Appendix 8) that incorporates probable (but not included in the variances) cost and time impacts, may be used to provide an early warning of possible cost risks. A means must be developed to reflect these trends into the cost forecast format. As a project progresses, forecast contingency may typically be reduced to reflect the increased level of certainty of committed costs. Notes A and B in Appendix 8 show the reduced contingency (reserve) for items as they are committed (1/2% vs. 3% reserve). In this example, the forecast contingency would be periodically re-evaluated by applying a lower percentage to committed and spent tasks, while retaining the higher percentage for uncommitted tasks.

The Cost Trend Log is a summary of known project cost risks that can be equated to forecast contingency (reserve) needed to complete the project. As such, it is a contingency variance.

Disputes between the owner and contractor, or contractor and subcontractor, may have less certainty in the cost estimate. The Cost Trend Log (Attachment 8) makes provisions for these potential trends by incorporating these less certain risks, Note C, via a Risk Management Log. Please note that the Risk Management Log will be discussed in Section 3.6. As can be seen from the example, if only the negative cost variance, Line 32 was considered, the contingency would be reduced by $347,955 without any determination if the remaining contingency $3,206,012 minus $347,955, or $2,858,057, would be available for completion. The Cost Trend Log summarizes the needed contingency based on completion status, as well as that from the Risk Management Log. Thus, the available contingency for completion is $1,719,410 instead.

While disputed scope definition and schedule delays affect project costs, change orders due to owner initiated scope changes or unanticipated field conditions are typically the most frequent contribution to cost overruns for an owner. Cost overruns and schedule delays apply equally to fixed cost and reimbursable cost contracts. Of course, timely resolution of all changes is essential to address the cost increase, especially where change order work impacts other ongoing work or also has a time impact. All change orders (whether between owner and contractor or contractor and subcontractor) should be tracked to identify how long a change order is in the process of approval. This "aging" information is useful for determining the need for management or other intervention.

Change order costs associated with significant redeployment of resources or delays can be evaluated if the project schedule is cost or resource loaded. The project schedule is a useful tool to assess the cost/time trade-offs for such changes.

Finally, for relatively larger change orders, an owner could obtain an independent estimate from a specialist estimating entity to validate the contract change proposal. This will facilitate a decision on both the method and amount of payment. If cost or time impacts cannot be agreed to, a *force account*, or time and materials tracking system are typically implemented to allow the work to proceed while discussions are ongoing. Alternately, a contract provision, discussed earlier, for an independent dispute arbitrator can expedite critical change order negotiations.

3.3 SCHEDULE CONTROL

Early and continuous planning and scheduling of a project allows for prompt detection of schedule impacts (variances) with sufficient time to explore a range of options to maintain the schedule. Scheduling the project from beginning to end derives the benefits of expending resources early to stay on or ahead of schedule, rather than expending even greater (due to inefficiencies in increased staffing) resources at the end of the project to try to catch up.

A comprehensive project schedule provides the ability to communicate the need for timely decisions and expenditures of resources. This is as true in the design phase as in the construction process.

In planning the construction phase, it is a lot easier to erase an activity on a schedule and redraw the sequence than to move people and equipment in the field. Too often there is not enough schedule definition of administrative activities, such as preparing and approving shop drawings or processing design clarifications via requests for information (RFI). As such, significant administrative activities including, submittal approval, fabrication and delivery sequence, as well as the related field storage and erection sequence activities could be an unrecognized impact.

As an example of resource options, Figure 3 shows a sequential or finish to start, (finish the entire preceding task before starting the next) schedule that satisfies the contract duration for installation of sewer piping. However, this is not the typical way the project is to be built. It is essential that the work is broken into tasks of reasonable work day

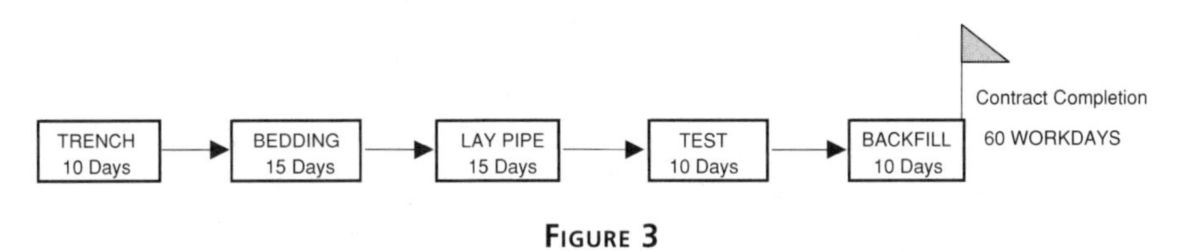

FIGURE 3

durations (15 days tasks, in this example). Tasks that 22 work days are difficult to manage unless subdivided into smaller tasks. A sequential plan often does not provide adequate flexibility to respond to delays. By subdividing the activities into concurrent (those that may occur simultaneously) tasks, the work can be completed in a shorter time with the same total resources. Figure 4 shows a possible approach, given manpower availability. Critical activities are related by logic constraints and durations. A delay in one activity will delay the project completion date. The critical (longest) path is determined by summing the longest duration for each sequence below. In essence, because of the shorter overall duration, Figure 4 builds in *total float* (the schedule equivalent of budget contingency) to the contract completion date, thereby providing for potential disruptions. This shows that projects can sometimes be accelerated by re-sequencing, as well as the obvious method of providing additional resources to reduce critical path durations.

One cannot afford to spend a great deal of management time and resources on activities that are not "critical" to project completion. On many projects, a small percentage of the schedule activities are critical. Because some of the "near critical" activities (those that are closest in terms of total float to critical path activities), not expected to be critical, could subsequently prove to be the bottlenecks that ruin any chance of on-time project completion, near critical activities should also be closely monitored.

Periodic schedule updating (at least monthly or when a significant issue is identified) allows all project team members to better plan and coordinate their work efforts. Similar to physical measurement of completion of the work for cost control, measurement of actual work completed vs. the comparable planned duration is essential for a meaningful schedule update. That is, completion of work must be based on physical units (feet of pipe, cubic yards of concrete, etc.) and not measured in terms of an estimated proportion of the planned duration. This will enable variances to critical or near critical path durations (delays) to be identified.

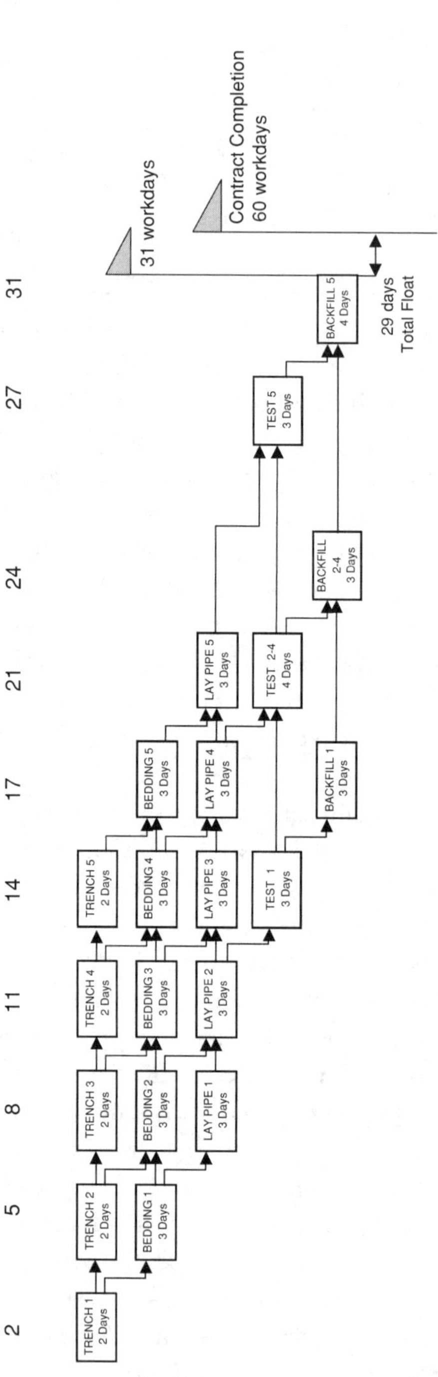

Cumulative Days:
2
5
8
11
14
17
21
24
27
31
TRENCH 1
2 Days
TRENCH 2
2 Days
TRENCH 3
2 Days
TRENCH 4
2 Days
TRENCH 5
2 Days
BEDDING 1
3 Days
BEDDING 2
3 Days
BEDDING 3
3 Days
BEDDING 4
3 Days
BEDDING 5
3 Days
LAY PIPE 1
3 Days
LAY PIPE 2
3 Days
LAY PIPE 3
3 Days
LAY PIPE 4
3 Days
LAY PIPE 5
3 Days
TEST 1
3 Days
TEST 2-4
4 Days
TEST 5
3 Days
BACKFILL 1
3 Days
BACKFILL
2-4
3 Days
BACKFILL 5
4 Days
31 workdays
Contract Completion
60 workdays
29 days
Total Float

Figure 4

Early knowledge of a delay will allow exploration of options such as:

- identifying those critical activities which could be shortened by the application of additional resources (i.e., increased staffing or selected overtime)
- identifying those activity sequences which constrain the completion date and then considering re-sequencing to eliminate or reduce the constraint
- accepting the reality of a delayed completion date and planning alternatives to the entire project being done on time (phased completion)

Figure 5 shows an example of Planned, Impacted, and Recovery Schedules for the "rough-in" phase of a project. Consider the initial schedule impact due to a survey error by the contractor. This error results in a two-month delay (design/build) in completion of the piling and an additional two-month foundation redesign impact. Thus, as shown on the Impacted Schedule, the contractor is responsible for a four-month

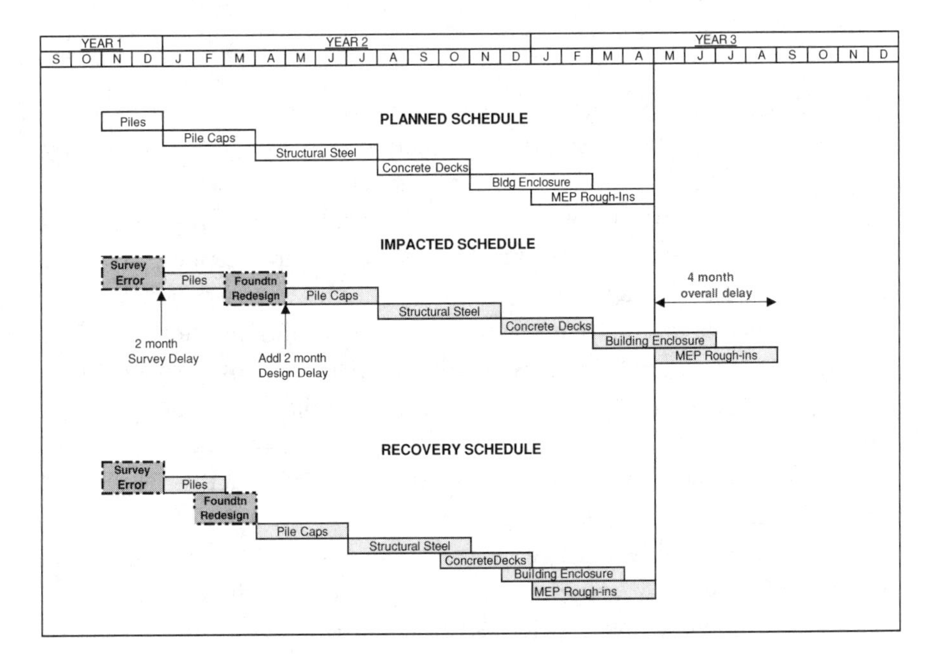

FIGURE 5

delay to the start of the pile caps, resulting in a four month overall delay. The Recovery Schedule shows where some recovery may be obtained by concurrently completing activities. By accelerating foundation redesign and concurrently performing concrete decks, building enclosure and Mechanical, Electrical and Plumbing (MEP) work, the net result is that the contractor has recovered the duration for completion of rough-in.

While the Recovery Schedule provides a means to reduce the time variance, the Planned Schedule must be maintained as the document of record, along with periodic updates, or as-built schedules. Comparing actual or physical completion with the planned completion (especially for partially completed activities) provides the basis for forecasting the overall project duration. The comparison should also identify the reason for the variance in an activity and which entity is responsible. This is necessary to evaluate compensation for the impact.

In determining the responsibility for a delay to project completion, and whose risk it may be, consider that there can be two basic types of delays on a project; excusable and nonexcusable. These terms apply to the contractor's performance requirement. Unusually severe weather and delayed access are examples of excusable delays to a contractor.

Excusable delays are often compensable by the owner and the contract time is extended, such that liquidated damages or other owner delay remedies do not apply to the extended contract duration.

Inadequate staffing of the project and rework by the contractor results in nonexcusable delays. Nonexcusable delays are not compensable and could expose the contractor to owner's delay related remedies.

If an excusable and a nonexcusable delay occur during the same time period, they are said to be concurrent delays. (Do not confuse concurrent tasks with concurrent delay responsibilities.) Typically, an extension of contract time is granted for concurrent delays (delay remedies do not apply) but no compensation is granted to the contractor. These are excusable but not compensable.

The contract should identify which delays are excusable as well as which are compensable. As an example, some contracts contain a No Damages for Delay language clause, which is intended to deny compensation for owner-caused delays by granting only a contract time extension. In this example the contractor may also bear full responsibility for concurrent delays. Thus, the delay is neither excusable nor compensable. Because there are limits to application of this clause, documentation of schedule related impacts is important for all entities.

3.4 SAFETY AND QUALITY ASSURANCE

As discussed earlier, safety and quality can have an impact on timing and cost. The selected contractor should have proven safety and quality assurance programs. Safety assurances shall include all relevant issues including those of the Occupational Safety and Health Administration (OSHA), industrial hygiene, property damage, third party protection, etc.

An owner can evaluate a contractor's safety program by requesting safety information from the contractor in the bid documents. These could include:

a. The contractor's commitment to safety, as demonstrated by an ongoing safety program that is supported by its top management
b. The completeness of the contractor's safety programs and their appropriateness for the work in concert with the safety standards of the owner and need to protect the public, where applicable
c. The contractor's response to safety questions included in the bid documents
d. The contractor's performance as compared to its industry peers

The use of independent safety audits is an important control tool. For renovation projects, a phasing plan with well-defined swing space needs, is an important safety/security risk identification tool as well as schedule tool.

Quality, in a like manner, is a function of the contractor's quality assurance program and owner's quality needs. Compliance with all applicable codes is a minimum quality requirement. The evaluation of the quality assurance program includes consideration of frequency and extent of quality control inspections by the regulatory and controlled inspection agencies, contractor and owner, as well as timely availability of agreed to quality control reports to the owner.

In some instances, quality "hold points" (concrete curing, etc.) may be included in construction schedules. These need to be clearly defined.

The level of enforcement of requirements for quality and/or safety can have a measurable effect on the success of a project. It can affect costs, schedule, and even project morale. It is, thus, important that highly visible steps be taken to effectively maintain safety and quality awareness for all entities.

3.5 INSURANCE AND BONDING

Insurance and bonding are tools used to transfer risk. A brief summary of those specific to design and construction are noted.

3.5.1 Insurance

Insurance requirements for a project include General Liability, Workmen's Compensation, Automotive, Builders All Risk, and Professional Liability (for a design professional). The owner typically specifies the amount of coverage required of the design professional and contractor.

One category of risk, the risk of having to rebuild completed work due to fire, storm or acts not under control of the contractors or owner, is typically managed via a "Builders All Risk" insurance policy. Other risks include providing sufficient liability limits and include coverage for third parties and adjacent properties that may suffer damages from unanticipated conditions during construction.

Another example, that is, errors and omissions risks, are ostensibly managed by designers' professional liability insurance. However, seeking this remedy means that the poor design has resulted in cost, time or quality issues. Rarely are owners adequately compensated for their losses. Instead, active involvement in the design and bid process will alert the owner to risks and allow additional resources to be involved in quality control of the designer's documents.

For modifications to an existing facility, an owner may also consider business interruption insurance.

Insurance does not cover delays due to strikes, labor lockouts or non-performance of key equipment.

3.5.2 Bonding

Types of bonds typically utilized include:

- Bid bond
- Payment bond
- Performance bond

The companies that provide these are called sureties. Basically, a surety bond is a guarantee. They are third party agreements between the surety, owner and the contractor, under which the contractor and surety guarantee the owner that the requirements will be completed. A surety bond is not an insurance policy. As stated, the surety is a guarantor, not an insurer. Surety bond premiums are not based on actuarial or statistical probabilities of loss. The underlying principle is that of a credit device.

In many instances, contractor performance cost risks are managed by the owner via performance and payment bonds for fixed price or GMP projects. (Similarly, a subcontractor performance risk is managed by a contractor with subcontractor bonds. This ensures that a surety will complete the work should the contractor fail to perform. In reality, the risks that are managed by bonds are only related to costs for a well-documented scope. To fully benefit by the use of bonds to control schedule (and thus performance) risks, an owner (or contractor) must have an effective early warning management system. As previously discussed, this includes a thorough understanding of scope definition (and the quality of bid documents), minimization of unknown conditions (via field investigations) and timely participation in schedule reviews. Because sureties are not typically active project participants, ongoing management by the owner, (or contractor) and early notification to the surety for negative trends is essential. However, the responsibility for the issues must objectively be weighed before notifying the surety.

3.6 RISK MANAGEMENT LOG

As stated above, the role of project management is to manage risks in design and construction. Project management is not the responsibility of an individual or the split responsibility of divided organizations. Rather it is the result of a cohesive team of owner, engineer and contractor. Understanding the initial risks and implementation of a continuous risk management process will improve the likelihood of successfully achieving the expected project performance goals.

Risk must be evaluated at each phase of a project, and must be effectively communicated. Appendix 9 is an example of a contractor's Risk Management Log for the Construction and Commissioning Phases of a design/build project. Each entity should have (and share) a log addressing respective risks. Once a risk is identified and quantified, it must be translated to a common basis (usually costs) or a critical success goal (cost or time). This provides the basis for prioritizing risks and establishing action plans.

Please note that the schedule risk, Item 2 of Appendix 9, has been quantified separately, in terms of time, because this delay results in a cost risk (in terms of liquidated damages or acceleration). Thus, this log shows the dependencies of various project goals, converted to a common cost basis to aid in establishing priorities.

The actions need to be tracked and expedited by periodic use of such a log. This log is not an isolated project document. It is part of a comprehensive trend report (as discussed in Section 3.2), providing an early warning of the risk to goals if attenuation methods are not successful. As discussed earlier, the cost of these risks would be addressed by application of available contingency.

The Risk Management Log also contains unique and ongoing issues. It also contains the unexpected (Item 3). Often, a contractor will work with an owner on a change in scope without final cost or schedule, even though the contract may require agreement before proceeding. A similar situation exists between a contractor and subcontractor. When this is the case, incorporating disputed amounts into the trending procedure would assist in evaluating both cash flow and total cost risks. Further, if there is insufficient trust or authority to expedite resolution, the probability is high that risks will be realized and disputes will become formalized as defined in the contract.

Item 7 on the Risk Management Log is a potential claim with some merit from a subcontractor. In the example, there is reasonable documentation to support the $75,000 risk. The claim amount must be quickly evaluated and incorporated so that actions (documentation, follow-up, etc.) may be prioritized. If it is not possible to do so by the project team independent outside input should be sought; collectively or individually. The Risk Management Log should not contain low probability or large contractor claims because it may greatly exceed the value of other items on the list and result in exaggeration of the trend on the Cost Trend Log. These should be addressed in an "off balance sheet" report by the affected entities as possible impacts. In any event, these claims require an immediate plan and application of appropriate technical or legal resources.

3.7 CONCLUSION

In any project, variance from the project plan is to be expected. All successful projects have procedures in place, at the beginning, to deal with anticipated variances. The most successful projects minimize variances by anticipating or evaluating potential risks and addressing them as early as possible.

In short, effective "continuous variance seeking" management is the first step in effective control. Timely analysis of negative variances and trends allows quantification of risks and comparison of the best available opportunities (not a single opportunity) for control, and therefore, success. It has been said that, "success has many fathers, but failure is an orphan." Should disputes not be addressed in a timely manner, and result in postconstruction claims, the team has produced many orphans.

Project "Right Start"

4.1 OVERVIEW

Having provided for a risk management process to reduce risks, incorporation at the start of a project, along with other planning tools, provides the framework for success! This will be referred to as the "right start" template for a project.

4.2 PARTNERING

As discussed earlier, owners expect a quality product, safely, on time and within budget. Engineers wish to participate in a challenging enterprise, make an equitable profit and satisfy the customer. The same is true for contractors. Therefore, a shared vision would incorporate these expectations.

Partnering is also an important component of managing risks and avoiding disputes. In order for contracting entities, such as owner and designer, or owner and contractor, to believe that a shared vision of success is important for an individual project and a long-term relationship, the element of trust is essential. Trust is also essential in risk sharing. However, trust cannot develop without the commitment of senior management and an acceptance of the "culture" and expectations of each partner.

Partnering sessions are conducted with the aid of an experienced facilitator. The Construction Industry Institute, Special Publication 17-1, In Search of Partnering Excellence, July 1991, Table 2.1[8] is repeated in Appendix 10. This table summarizes the key differences in approach between projects done with partnering practices vs. those done with the more common or "traditional" methods. Some of the differences, such as free organizational access, require much effort.

Partnering is an outgrowth of the manufacturing quality control movement after World War II. Dr. Edwards Deming was a leader in the methodology to "do things right the first time, and focus on continuous improvement." This methodology consisted of 14 points[9] (see Appendix 11). Many of the 14 points have been applied to service and construction businesses for quality assurance and are implied as the basis for trust in partnering. Some of the items, such as numbers 4 and 9, may have limitations in public projects and those with union labor, respectively.

The International Standards Organization's Standard 9000 is an example of a quality assurance standard that has been adopted in the design and construction service industry, fostering improvement in delivery of services. It has been considered a framework for a partners expectation, that of design quality assurance.

4.3 RESPONSIBILITY MATRIX

The breakdown in trust between parties is often the result of misunderstanding. While contract clauses often discuss the "what" and "how" of a provision, the "when" and "who" may not be specific. A change order clause, for example, may be silent on who in the organization is authorized to approve various dollar amounts or extensions of time and how long the various approvals should take. Consider establishing such a matrix for change orders as follows:

Change Orders	Responsibility, Extension of Time	Responsibility		Duration of Approval, Days	
		Under $50,000	Over $50,000	Under $50,000	Over $50,000
Owner	N.G. Way	R. Smith	F. Jones	10	30
Designer	J. Green	J. Green	M. Gray	15	15
Contractor	L. Rich	L. Rich	G. Glass	15	15

Often, the responsible person for the designer and contractor are the same for all approvals. However, the owner, particularly in the public sector, may have restrictions on amounts approved at the project level in the field or wish to use an independent estimator for large change orders or those requiring an extension of time.

A similar matrix could be developed for resolution of scope disputes that could lead to a change order.

Of course, a matrix is also applicable for a contractor to ensure timely coordination with subcontractors on similar issues. However, the contractor may be constrained (i.e., timing of owner reimbursement) in dealings with a subcontractor.

4.4 PROJECT PROCEDURES MANUAL

Many companies have project procedures manuals. However, they are often general and not specific to a project. An example is the close-out (or transfer) documentation. As soon as a contractor is awarded a job, the close-out file should be indexed. Specifically, what test and inspections will be required for a Certificate of Occupancy or fit for service declaration and which regulatory agencies are involved?

Topics to be included in a procedures manual include:

1. Close Out Requirements (required for a Certificate of Occupancy, or fit for service declaration)
2. Organization Chart and Contract Information
3. Definition of Contract Terms (i.e., substantial completion)
4. Partnering: identification of individual risks and cooperation expected
5. Responsibility Matrix
6. Project Risk Management Plan (for deviations in scope, cost or time)
7. Payment Procedures and Documentation
8. Request for Information process
9. Change Notification process
10. Change Order, or Credit, process
11. Backcharge Authorization Process
12. Cost Tracking and Forecasting
13. Scheduling Requirements and Access Constraints
14. Meetings and Responsibilities for documentation (i.e., writing meeting notes, two-week look-aheads, etc.)
15. Monthly (to Owner) Reporting Requirements

16. Submittal Process
17. Procurement Tracking and Expediting
18. Safety and Security Assurance and Reporting
19. Quality Assurance (including Notice Requirements for Controlled inspections) and reporting
20. Regulatory Requirements
21. Utility Applications
22. Photographic Requirements
23. Construction Permits
24. General Conditions
25. As-Built Requirements
26. Warranties and Guarantees
27. Commissioning or Turnover Schedule Requirements

4.5 TEAM EFFECTIVENESS

Partnering principles are intended to develop effective teams. As a project progresses, strains relating to competing goals develop. As in continuing to improve project performance goals, improvement in relationships is of equal importance to the continued success of the project. In some projects, periodic partnering sessions are utilized to maintain mutually beneficial working relationships. An alternate technique utilized to facilitate continuous improvement in the effectiveness of the project team is the "Team Process Check." This technique, utilized frequently in the manufacturing sector, is typically utilized without the aid of an outside facilitator. The following is an example of a Team Process Check conducted at the end of quarterly construction meetings. Management review of the results is essential.

1. Team rules (no cell phones, handouts prepared in advance, etc.) should be established and posted.
2. Each meeting should have a team leader, recorder and a timekeeper to maintain focus. The timekeeper signals the group to do a team process check approximately 15 minutes before the announced finish time.

a. Rate Each Dimension.
   Each individual privately writes down a rating for each dimension (see Table 1).
b. Record the Ratings.
   The recorder asks each person (round robin) for a rating on each attribute until all the attributes have been rated. Every person's answer is recorded on the flip chart. (Mark or flag extreme ratings for later discussion.)
c. Compute the average (x) and sample standard deviation (s) for each dimension, and plot a histogram. These will be used for reference after the next team check. Remember, the average will show the center of the data set, and the standard deviation* is the measure of variation of the data.

$$*S^2 = \frac{\text{Sum(data point} - \text{x})^2}{\text{Number of data points minus one}}$$

d. Discuss the Reasons for Ratings.
   The team leader asks for individuals to explain their ratings. The purpose is to give individuals and/or the team feedback. It is a good idea to begin with positive feedback and work toward more negative issues. It helps to focus on:
   (1) Specific behaviors. What particular actions led to a given rating? Be descriptive—not evaluative.
   (2) Impact on working relationships. What reaction did the behavior cause (e.g., feelings or perceptions)?
   The discussion should also focus on any high and low ratings on a given attribute and any high or low averages or standard deviations.
e. Keep a Continuous Improvement List.
   Ask team members to suggest ways to improve, and list the items on the flip chart. Examine the list for items on which the team agrees could improve the working relationships and decision quality in subsequent team meetings.

Table 2 is an example of the scores received at a sample meeting.

TABLE 1. Team Effectiveness Measurement Process Check

Circle the number on each dimension which reflects your rating of the team.

ON-TRACK

1	2	3	4	5	6	7	8	9	10

NO AGENDA OR DID NOT FOLLOW THE AGENDA — FOLLOWED THE AGENDA - NO DIGRESSIONS

PARTICIPATION

1	2	3	4	5	6	7	8	9	10

A FEW KEY MEMBERS DOMINATING AND SOME MEMBERS NOT PARTICIPATING — EVERYONE CONTRIBUTES AND IS INVOLVED IN TEAM DISCUSSIONS

LISTENING

1	2	3	4	5	6	7	8	9	10

MORE THAN ONE PERSON TALKS AT A TIME: REPETITIONS, INTERRUPTIONS, AND SIDE CONVERSATIONS — ONE PERSON TALKS AT A TIME: CLARIFYING AND BUILDING ON IDEAS

LEADERSHIP

1	2	3	4	5	6	7	8	9	10

OWNERSHIP/RESPONSIBILITY DILUTED BY DELEGATION — OWNERSHIP/RESPONSIBILITY DEMONSTRATED BY ENTIRE TEAM

CONSENSUS BUILDING

1	2	3	4	5	6	7	8	9	10

TEAM BEHAVIOR WAS CONFLICTING WITH CONSENSUS BUILDING — TEAM BEHAVIOR WAS ASSISTING CONSENSUS BUILDING

DECISION QUALITY

1	2	3	4	5	6	7	8	9	10

TEAM DECISIONS WERE NOT TIMELY OR WERE DISPUTED — TEAM DECISIONS WERE TIMELY AND NOT DISPUTED

TABLE 2. Meeting Example

Name	On-track	Partici-pation	Listening	Leader-ship	Consensus Building	Decision Quality
J. A.	10	7	9	6	9	8
B. A.	8	6	7	8	7	7
P.B.	8	6	8	7	8	7
U.B.	8	3	6	7	8	8
J.E.	9	7	8	7	8	8
N.E.	8	6	8	7	7	6
L.H.	10	7	9	8	6	4
S.J.	9	7	7	8	8	8
D.K.	8	6	8	6	6	7
J.M.	8	6	8	7	8	8
E.M.	9	6	7	7	7	8
R.N.	9	6	8	9	7	7
M.O.	9	7	9	9	8	8
L.P.	8	7	7	9	7	8
R.P.	8	5	7	6	NA	NA
K.P.	10	7	7	7	5	5
R.S.	9	7	9	9	8	8
L.S.	9	7	7	8	6	7
M.S.	8	7	9	9	7	9
D.S.	8	5	7	7	5	6
B.S.	8	7	8	9	8	8
A.P.	9	6	8	8	8	8
Average (X)	8.64	6.27	7.77	7.64	7.19	7.29
Standard Deviation(s)	0.73	0.98	0.87	1.05	1.08	1.19

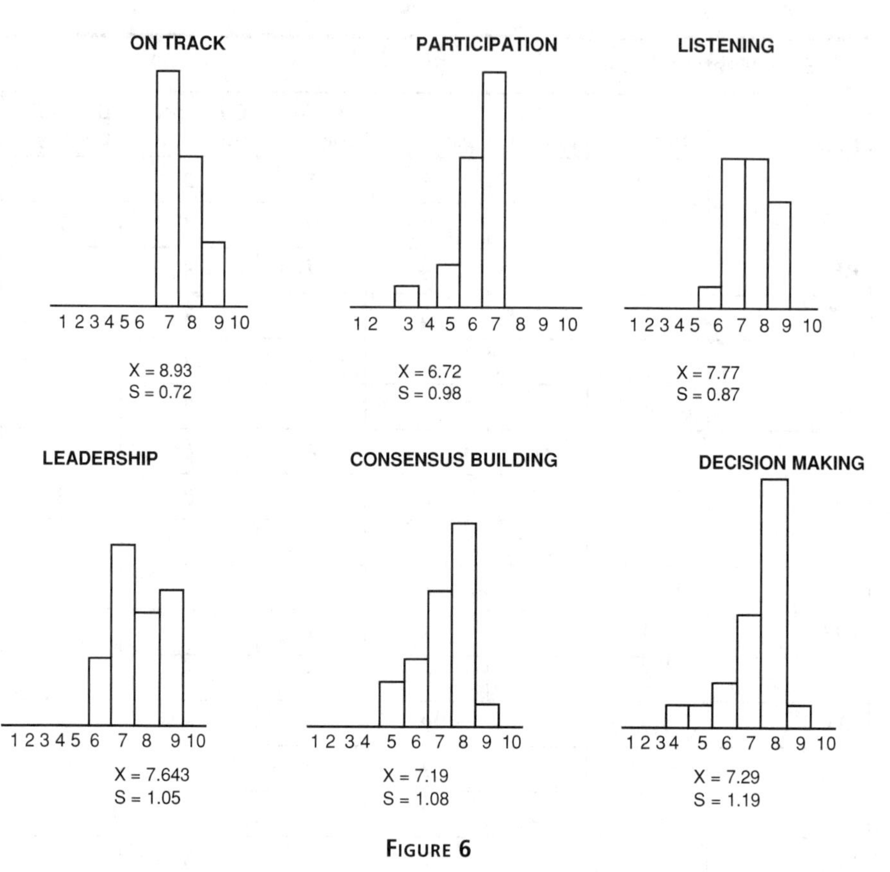

Figure 6

The information from Table 2 is summarized as follows:

1. On Track	X = 8.64,	S = 0.73
2. Participation	X = 6.27,	S = 0.98
3. Listening	X = 7.77,	S = 0.87
4. Leadership	X = 7.64,	S = 1.05
5. Consensus Building	X = 7.19,	S = 1.08
6. Decision Making	X = 7.29,	S = 1.19

Figure 6 graphically illustrates the results. The lowest average score identifies the need to improve Participation. Examples include:

- Please avoid telephone interruptions or walking out of the room during the meeting (team rules that prohibit these and other actions may be modified, if not included).

- Participation does not necessarily mean asking questions. Being present and attentive is also a way of participating. Do not participate in side discussions.
- Lack of participation negatively impacts other dimensions.

The largest standard deviation, for Decision Making, implies a large variation in the decision-making consensus. The techniques applicable to improving team decisions need to be explored.

4.6 PROJECT HEALTH ASSESSMENT

Each project has requirements for reporting on the project goals both to the owner and to the contractor management. As discussed, these goals and risks must be quantified for effective management oversight. The objective of a summary report is to succinctly present the project status, or its *health*, by comparison with the baseline goals of the project. A summary report is easily read by management, especially where many projects are ongoing simultaneously. Figure 7 is an example of an owner's one-page report.

Of course, this health assessment does not replace the more detailed reports needed by the organization to support the status presented. The more comprehensive report would include financial details and summaries of the project control tools discussed in previous chapters, short-term achievements and plans, photographs and narratives on successes, opportunities and failures (the so called: Good, Bad, and Ugly).

4.7 OVERSIGHT

In addition to establishing a good project implementation plan, periodic review, or oversight, should be considered an element of the risk management process. On-site project personnel can become overwhelmed in the details of construction and, thus, delay risk management activities. Often this results in inaccurate reporting and a loss of capability for timely response.

PROJECT HEALTH ASSESSMENT REPORT

Contractor: **Contr.No.** **Date:**

FINANCIAL ASSESSMENT

Budget ________ ***SOURCE/COMMENT***

Forecast to Complete________

Available Contingency________

Timeliness of Progress Payments

Acceptable ________ **Concern** ________

Last Payment Record:

Date "Pencil Copy"Approved ____

Date "Hard Copy" Submitted to Finance ____

Payment Date (Vendor Date) ____

Timeliness of CO Processing

Acceptable ________ **Concern** ________

C.O. Data:

Executed COs________ Amount ________

Pending COs________ Amount ________ Max.Time Pending____

SCHEDULING/COORDINATION ASSESSMENT

Adherence to Approved Phasing Plan ***SOURCE/COMMENT***

Acceptable ________ **Concern** ________

Next Major Milestone ________

Milestone Constraint/Action ____

Substantial Compl. Date *Orig.* ________ *Forecast*________

Percent Complete *Planned*________ *Actual*________

Lowest Total Float ________

SAFETY/SECURITY ASSESSMENT

Adherence to Site Safety/Security Requirements ***SOURCE/COMMENT***

Acceptable ________ **Concern** ________

Safety/Security Data When Concern Noted:

Dust/Air Pollution Provisions Acceptable ____ Concern ____

Noise Provisions Acceptable ____ Concern ____

Work Area Segregation/Security Acceptable ____ Concern ____

Sidewalk Shed Maintenance Acceptable ____ Concern ____

Safety Recommendations Report Date: ________

Safety Incidents To Date ________

Recommendations New ________ Repeat ____

Hazardous Mat'l Incidents to Date ________

QUALITY ASSESSMENT

Adequacy of Contractor's QC Program ***SOURCE/COMMENT***

Acceptable ________ **Concern** ________

Overall Rating by Discipline **M E PI R S GC**

Inspection Report Date

Const. Deficiencies this Report New____ Repeat: ____

Const. Deficiencies to Date ________

Deficiencies Remining Open ________

FIGURE 7

Oversight is not an audit. Audits are postmortems and deal with performance in the past. They carry an undertone of criticism and punitive connotation. Oversight teams use past performance to learn from, and to encourage future excellence while discouraging poor performance. What is going right and why, is just as important as what is going wrong and why. Through observations of past performance and projections for future performance, project oversight may result in recommendations and a comprehensive action plan to implement improvement opportunities. The objective is to assist in meeting or exceeding all project goals.

Operating companies often conduct formal reviews prior to construction funding approval by Boards of Directors. This "design review" focuses on the elements of risk identified previously (contract type, delivery method, completeness of scope, definition, budget, cost contingency, schedule, schedule contingency, and planned resources). This effort is intended to support that of the project team's and confirm the project goals are consistent with corporate goals. An example of a Project Oversight Checklist for a preconstruction review is given in Appendix 12. The items on the checklist are indicators, based on prior experience. For example, Item 1.10 would indicate how much design definition has gone into the project to date. Experience would allow determination of the potential risks associated with the estimated cost and timing.

Similarly, periodic (quarterly, etc.) oversight of the Construction Phase is often practiced by operating companies and those that have a financial stake in the project. This provides an independent assessment of the status of the project performance vis-à-vis the goals, and associated risks, and identifies best practices and lessons learned to share with other projects. Appendix 13, Risk Memos,[10] is an example of a financial oversight tool utilized by large contractors to provide internal assurance that all risks are being addressed. Please note that time related and other risks must be converted to a cost basis to complete the required element III A7.

Appendix 14 is a process diagram for oversight of a project of a long-term capital program. Depending upon the project size and participants (joint venture, etc.), the composition of the oversight team will vary.

While the focus of the oversight team is on the business and management issues essential to project success, a process or product expert may be part of the team when the technology is "substantially" different from previous projects. Nevertheless, the smaller the oversight team, the more efficient are the reviews, resulting in less impact on the project staff's day-to-day activities.

Resolving Project Disputes and Claims

5.1 OVERVIEW

Application of an effective risk management process will substantially increase the probability of success and reduce the probability of claims. It is true that an owner, contractor, etc., may resist having to pay for a specific issue (extra work, delay, etc.) but more often disputes are due to poor communication and lack of understanding of the issues involved. Since "poor communication" not only includes clear identification of the issues (that are part of the risk management process), but an understanding of the expectations of each party, some points that have contributed to prevention or resolution of disputes will be addressed.

A claim has two components—entitlement and damages. Entitlement is identified by the project tools discussed previously and is presented in summary fashion in this chapter.

Damages are calculated by contract provisions or legal precedent. These will be discussed.

The previous discussions enumerated the process or procedures necessary to prevent or resolve disputes. The following sections will explore these in more detail.

5.2 CONTRACT PROVISIONS

Control of risks and claim prevention begins with a thorough understanding of, and compliance with, key contract provisions and procedures that address them. Therefore, project managers or lead personnel must be familiar with all contracts and subcontracts. For example, before changes to a contract cost or duration can be authorized, agreement on procedure for quantification is necessary. Since subcontractors are often bound by these provisions, a copy of the prime contract provisions should be provided to subcontractors. Typical provisions include:

1. Substantial/Final Completion: Project completion should be well defined. In some instances substantial completion and beneficial use are interchanged. However, if a project is being beneficially utilized, the project may be partially substantially complete. Thus a clear definition of punch list-type activities that can be done after substantial completion should be included. Reduction of retainage after substantial completion should be addressed. A time period should be defined for completing the work after substantial completion and providing the documentation required for final completion.

2. Liquidated/Other Damages: Contracts contain provisions for liquidated damages or provisions for consequential damages when a contractor is late in achieving substantial completion (usually). Some contracts include provisions for no (owner) damages for delay. Because delay damages may be significant, and are a major cause of claims, the importance of continually evaluating schedule risks and timely notification of impacts cannot be overstated. Most contracts require notification of any delay to the overall project completion date or milestone dates. Time frames for notification range from 24 hours to 30 days. It is essential that this be complied with so that the owner may mitigate the impact, if possible.

3. Differing Site Conditions, Changed Conditions Or Concealed Conditions: Many construction contracts have some form of provision to cover the unexpected which is experienced when excavating a site or opening an existing structure. Because these are different from extra work that typically improves the project, unit prices for potential activities should be considered. For example, if soil borings indicate a certain rock type, provisions could be made for the possibility of discovering a different or more extensive rock. Clear notification requirements and documentation of suitable test results for this type of change, as well as timely notification are implicit.

4. Authority Of Parties: The contract should clearly spell out the authority of the parties to approve change orders and provide for as much authority at the project level as possible. In addition to publishing an organizational chart with names and titles stating responsibilities, it should be clear what change order amount can be approved by each owner's representative and how long the authorization process should take. This is essential for prompt payment for earned work. If this is not in the contract, a Responsibility Matrix (see Section 4.3) should be created at the start of a project.

5. Variation In Quantities: The standard clause in many unit price contracts simply states that if there's a variation in quantities, plus or minus a certain percentage, both parties will agree to negotiate a new price. This does not provide for a resolution of the problem. Equitable contracts specify a pricing structure for changes for various ranges of the quantity of work required.

6. Rates And Markup For Time & Material (T&M) Work: There is always a possibility that a change in the scope will require work to be performed on a T&M, or cost-plus basis (sometimes called a force account). Therefore, it is important that the contract include not only a clear statement of the rates (or the basis for establishing these rates) to be used for labor and equipment, but also the markup for overhead and profit to be applied to labor, equipment, material and subcontract work.

7. Resolution of Disputes: Most construction contracts clearly state a limitation on when the contractor must submit its claim. It may be within "X" number of days from when a change order was rejected, or number of days of completing the work (which can be vague) or prior to acceptance of final payment. This must be clearly understood and strictly complied with. If a dispute cannot be negotiated, the two traditional possibilities for resolution are arbitration and litigation (court). Either may be preceded by mediation. There are pros and cons to arbitration vs. litigation. The same provision is needed with all parties on the project (multiprime projects). All too often disputes involve a number of parties on the same project while a common issue can only be resolved in a single forum.

8. Exculpatory Clauses: Contract exculpatory clauses are used in an attempt to shift the financial risks for a specified problem from one party, usually the owner, to another party. The presence of exculpatory language can create an adversarial relationship at the start of a job. A real example is:

> The Owner shall have made available to the Contractor, before the submission by the Contractor of the Tender, such data on hydrological and sub-surface conditions as have been obtained by or on behalf of the Owner for investigations undertaken relevant to the Works but the Contractor shall be responsible for his own interpretation thereof. The Contractor shall be deemed to have inspected and examined the Site and its surroundings and information available in connection therewith and to have satisfied himself (so far as is practicable, having

regard to conditions of cost and time) before submitting his Tender, as to:

(a) the form and nature thereof, including the sub-surface conditions,
(b) the hydrological and climatic conditions.
(c) The extent and nature of work and materials necessary for the execution and completion of the Works and the remedying of any defects therein, and
(d) The means of access to the Site and the accommodation he may require, and, in general, shall be deemed to have obtained all necessary information, subject as above mentioned, as to risks, contingencies and all other circumstances which may influence or affect his Tender.

The Contractor shall be deemed to have based his Tender on the data made available by the Owner and on his own inspection and examination, all as aforementioned.

9. Ambiguous Contract Language. It is probably impossible to prepare a contract with absolutely no ambiguity. Words like: reasonable, timely, prompt, workmanlike, etc. need to be looked at carefully to see if there is a replacement quantification phrase which will not be open to interpretation by both parties. A real example of both ambiguous and exculpatory language is:

> It is the intent of the contract documents to describe a complete Project to be constructed in accordance with the Contract Documents. Any Work, materials or equipment that may be reasonably inferred from the Contract Documents as being required to provide the intended result shall be supplied whether or not specifically called for.

5.3 CAUSES OF TROUBLED PROJECTS

If the contract provisions of the previous section are not addressed in your contract or, if the provisions have not been followed as stipulated or, if there are substantial design changes impacting the project, or if access is different from planned, the dispute probably cannot be quickly resolved. This section describes typical areas of disputes that are not easily resolved, "the usual suspects". This is an adaptation of the "The Deadly Dozen."[12]

1. *Errors and Omissions.* An engineer is responsible to provide a reasonable standard of care, such that the necessary field investigations and as-built drawings are adequately incorporated in the design documents. These documents must be complete and relatively free of errors or omissions. Errors that result in rework in the construction phase or omissions identified during construction may have impact on the cost and time to complete. Adequate professional liability insurance, typically specified in the contract, may be inadequate.

An important issue related to this is the objectivity of the design professionals such that equitable payment terms may be agreed with the contractor. See 10 below.

2. *Delays.* By definition, a delay results in completing the work which increases the overall project duration. The additional financing costs and the possibility of cancelled leases can be a hardship for an owner. Delays can also be expensive for contractors who face a longer period of project and home office overhead, as well as escalation of labor and material costs. Often, when delays occur it is no simple task to determine the responsible party. Each party will argue the delay was the fault of the other. An agreed to methodology for evaluating the impact and responsibility should be included in the contract. Please note that delays to noncritical path activities are really disruptions to the plan and are not relevant in this regard.

3. *Acceleration.* If an owner requests the work be expedited in advance of the baseline schedule or to mitigate delays, the contractor is entitled to *directed* acceleration costs. When a contractor falls behind schedule due to his own (self-inflicted) delay, the contractor should advise how he intends to get back on schedule (recover). An owner should not give the contractor any directives on how to recover the lost time. Otherwise the contractor may believe they are being directed and may be entitled to obtain acceleration costs. *Constructive* acceleration occurs when time extensions, which are due the contractor because of delays beyond his control, are not promptly granted. It is important to promptly respond to all time extension requests that are made by the contractor. If the contractor's request is not properly supported with CPM backup, more information should be requested/provided.

4. *Changed start date for construction.* The impact on a contractor's cost, when the owner shifts the contract start date, must be considered before awarding a contract. For example, a project that would have been constructed over two summers and one winter may instead have to be performed during one summer and two winters, and can result in an increase in construction cost. If the change in start date is addressed prior to signing the contract, the parties can negotiate the cost of the new start dated and there should be no claim. However, if the date is changed after the contract has been executed, it may be difficult for the parties to agree on the cost of the change. Refer to types of acceleration discussed in 3.

5. *Changed Work Sequence.* If the owner has to impose any restrictions on the sequence, the restrictions should be clearly stated in the contract documents. For example, in a renovation project close coordination with the occupants is necessary to identify any restrictions in sequence. If any restrictions are required, the restrictions must be addressed prior to contract award.

6. *Access Restrictions.* Access restrictions are an important subcategory of changed work sequence that have been discussed above. Any access restrictions such as limited work areas or business hour restrictions, should be clearly stated in the contract documents. While broader than Changed Work Sequence, when dealing with an operating facility, all potential bidders should attend a site tour during operating hours. This will help to eliminate expensive surprises.

7. *Excessive Management (Interference) by the Owner.* The construction contract should clearly define what the end product should be. Once that is done, the means and methods should be left to the contractor. The owner or its designer must not give direction to (interfere with) the contractor, or worse yet, give directions to subcontractors regarding alternate means and methods to be used. A single point of contact with the contractor, in accordance with a contractual organizational chart, is essential to reduce the potential for interference.

8. *Lack of Management by the Owner.* Very often an owner contracts with more than one party to accomplish the design and construction of a project. In most cases the owner hires an engineer or architect to design and a contractor to construct. The management of these two functions and their interfaces is clearly the responsibility of the owner. In other cases, an owner may hire two or more contractors (multiprime contracting) to build a single project. In that situation, the owner has the responsibility to ensure coordination of the separate prime contractors (Wicks Law) and may need to address the multi-prime arrangement with special provisions in the contract and provide an individual or a group that is qualified to play a proactive role in coordinating the work plans of others.

9. *Variation in Quantities.* It is not uncommon for an owner to embark on a project where it is difficult to determine with certainty the quantity of certain aspects of the work. For example, the engineering site investigation reveals there are large boulders in the area to be excavated but until the work is performed it is not possible to determine the exact quantity. In this situation, it is reasonable for an owner to use some type of unit price arrangement. Unit prices can be included in both reimbursable or fixed price contracts. In unit price procurement, the owner provides an estimated quantity of work to be performed as a basis for evaluating the bids, and both parties agree that the contractor will be paid a predetermined price for each unit of work performed. Unit price contracts work well when the actual quantities of work performed are fairly close to the estimated quantities; however, when actual quantities differ drastically from estimated quantities, inequities occur. For example, if a contractor mobilizes for drilling and blasting to excavate an estimated 1,000 cubic yards of rock and only finds 50 cubic yards, the unit price payments will not equitably compensate the contractor for the cost of mobilizing. The opposite can be true also when actual quantities far exceed estimated quantities.

If quantity variation clauses are not in the contract, these potential inequities must be dealt with as soon as possible.

10. *Lack of Objectivity for Change Orders.* A stream of changes to the project can affect the contractor's ability to complete the project on schedule and make a profit. The time to make changes in the design is during the planning and design phases, not during the construction phase. Changes required during the construction phase should be objectively evaluated and done as quickly as possible. They must also address both the cost and payment timing for the contractor as well as the impact on the contractor's schedule. Since errors and omissions in design documents are a major source of dispute, the owner may need independent technical advise to resolve a dispute. See 1 above.

11. *Failure to Perform.* Performance may mean adherence to the project schedule or scope related issues. Contractor delays are usually a material breach of a fixed price contract.

The owner has two clear-cut options when it comes to specifying the required end product of a construction project. The owner can hire a designer who will then prepare detailed plans and specifications describing exactly what the work should consist of, or the owner can write a *performance* contract stating the required performance and leave the detailed design to the contractor (design/build).

Problems arise when both detailed design and performance requirements are specified and are incompatible. For example, a contractor may be required to install an owner specified piece of air conditioning equipment and design the ductwork and registers to insure that the system will keep the facility at specified conditions. If these are not achieved, who is responsible?

5.4 COSTS ASSOCIATED WITH DELAYS

Delays can result in:

1. Extended project duration—resulting from delayed completion of the project.
2. Acceleration—to recover delays to the critical path (which would delay the project completion date), or due to reduction in the critical path contract duration during the course of the project.

Delay or acceleration analysis, which evaluate the responsibility for various tasks, is performed via CPM schedules, job meeting minutes, daily logs, submittal logs, and RFIs. This analysis establishes the entitlement—who is responsible and for what duration?

Since delays are the most difficult type of claim to analyze, and agree to, disputes are not uncommon. It is also true that this type of claim can have a significant impact on the economic benefit of each entity. Some entitlements include:

- **Owner costs associated with delay**
  - Loss of rental opportunity
  - Loss of manufacturing opportunity

- Loss of selling opportunity
- Additional costs for architect/engineer
- Additional costs for project management and inspection
- Additional finance costs

Delay damages for an owner may be limited to the liquidated damages specified in the contract. If none are specified, consequential damages may be applicable.

- **Contractor costs associated with delays**
  - Extended job supervision and field overhead
  - Extended equipment costs
  - Wage escalation
  - Material escalation
  - Extended home office overhead
  - Productivity losses
  - Unanticipated overtime
  - Finance costs
  - Reduced job opportunities/bonding restrictions
  - Lost profit margin
  - Loss of bonus, where applicable

Delay damages and loss of productivity are two significant sources of claims. Loss of productivity may not result from a delay to the critical path. Further discussion is warranted.

- Delay damages for a contractor are related to the nature and length of the delay. Time related costs for home office overhead are evaluated via methods such as the *Manshul Method*. This name is derived from a New York court decision, 79 A.D. 2d 383. Appendix 15 is an example of the Manshul Method. This method is intended to avoid duplication of overhead that occurs in other methods and to apply profit only to the additional overhead.
- Extended field costs and escalation costs for extended duration are fairly straightforward to identify. They may be supported by actual invoices or labor agreements.

- *Inefficiency* costs, due to acceleration, or productivity losses, are supported by actual costs and/or analyzed by techniques such as the *Measured Mile*. The measured mile approach compares actual productivity in an unimpacted period to that in an impacted period.

These costs may be due to a project delay, acceleration, changes in sequencing or other disruptions. They may apply even if acceleration is notsuccessful. Appendix 16, Acceleration model: Productivity Issues, is a modified version from the Construction Industry Institute's "Productivity Measurement: An Introduction", Publication 2–3, 1990.[13] It is a pictorial presentation of elements that impact productivity. It shows manpower allocation, work environment and other issues. As can be seen, acceleration often results in a combination of manpower allocation issues that contribute to a loss in productivity. Since productivity relates to standard work hours, unanticipated overtime premium costs are not included in a productivity analysis.

While the Measured Mile utilizes actual productivity information in a period that is not impacted by disruption, out-of-sequence, inefficient operations, stacking of trades, etc., and comparable information in periods with identifiable impacts, caution is in order. One must be sure the impacts being analyzed are the responsibility of one party, thereby giving entitlement to the other. Thus if a contractor claims entitlement for loss of productivity by impacts beyond their control, the contractor should support no contributing impact within their control. There are several Associations that have produced publications that list impacts affecting labor productivity. These include the Mechanical Contractors Association of America, the National Electrical Contractors Association, and others.

Productivity can be defined as:

$$\text{Productivity} = \frac{\text{Actual Output}}{\text{Actual Input}}; \quad \text{on a directly related basis.}$$

$$= \frac{\text{Work Produced in Units}}{\text{Hours Utilized or Cost Incurred}}$$

$$= \frac{\$\ \text{Revenues}}{\text{Actual Cost}}$$

$$= \frac{\text{Hours Earned}}{\text{Hours Worked}}$$

Often the inverse of productivity is utilized. This, Unit Rate, provides a basis to focus on improving productivity by optimizing hours worked for a given (actual) completion for various labor-related tasks.

$$\text{Unit Rate} = \frac{\text{Hours Worked}}{\text{Percent Complete}}$$

In manufacturing, output and input are routinely subdivided into measurable quantities. In construction, however, multiple sources of services and materials are involved, so output is difficult to associate with identified input. Thus in order to improve productivity, the total work hours used must be reduced. It is necessary to identify components that are directly related to the output and those which are not. Thus, the total work hours may be shown as:

$$\text{Total Work Hours} = X_1 + X_2 + X_3 + X_4 + X_5$$

Where: X_1 = Direct Hours

X_2 = Support or Nonworking Supervisory Hours (crew size dependent)

X_3 = Idle/nonproductive Hours

X_4 = Rework Hours, engineering error (or change order)

X_5 = Rework hours, contractor error

Productivity measures are typically $/workhour invoiced or % complete/workhour invoiced. The use of budget man-hours for productivity analysis is not appropriate because of the uncertainty in the validity of the bid estimates (could have been underbid), unless validated. Similarly, the use of requisition Schedule Of Values (SOV) is not appropriate either because of the undefined split between material and labor and undefined overhead and profit contained in various items. (The SOV is for payment purposes only and may have little relationship to the actual split of costs). As such, they would need to be subdivided to show the value for labor. Also, the use of "front end loading" (higher requisition value for early activities, as compared to the value based on actual work completed) is sometimes permitted in a SOV to improve a contractor's cash flow. Because this reduces the effectiveness of utilizing requisition information, time and material tickets or daily reports, which detail work performed, are the most useful support information.

Appendix 17 shows an example of the Measured Mile. The productivity issues are reflected in a comparison of actual productivity before and during the acceleration period. This is for similar tasks. A breakdown of these numbers to develop a Measured Mile analysis is shown for the bid and actual durations. The bid expectation is shown as a reference to determine if the bid was realistic. As can be seen, prior to and at the end of the acceleration period (April and May and October, respectively), performance measured by $Earned/work hour, was above the bid productivity expectation. The bid productivity was $113/work hour and the non-acceleration period exceeded $142/work hour. Thus, this bid was realistic. Also, the Attachment shows the productivity dropped to $50/work hour during the acceleration period. This shows that while the bid productivity was being exceeded and productivity was better than forecast during the height of the acceleration period (June through September), the performance averaged nearly half of the bid productivity. This could be directly attributable to the poor working conditions resulting from the acceleration activity.

5.5 CLAIMS RESOLUTION

There are many actions that the owner, designer and constructor can take during the life of the project to limit the possibility of claims and disputes developing. The previous discussions have addressed those actions. However, despite your best efforts, there will be times when two parties have different viewpoints on an issue and a claim develops. Because this is not uncommon, contract language typically addresses notification requirements and methods of resolution. Some methods include:

Mediation—This type is a more formal type of negotiation where a neutral third party (the mediator) attempts to facilitate discussion between the parties to ensure that the strengths and weaknesses of each party's position are discussed. Mediators have a variety of techniques that are used to foster a settlement. This is typically nonbinding.

Dispute Review Board—This technique is often utilized on large, long-term heavy construction projects such as tunnels, dams and highways. A panel of one or three individuals are chosen by the parties at the beginning of the project and are named in the contract to be the Dispute Review Board. This board, which is paid by the parties, monitors the progress of the project and can be mobilized, usually on a few days notice, to hold a hearing on a dispute and then render a decision which is binding on the parties (if specified in the contract).

Arbitration—This technique is similar to the dispute review board except that the arbitration panel is named after the dispute arises. There are a variety of methods in place for selecting the arbitration panel. The most common contract language is for the party who wants to initiate the arbitration to contact the American Arbitration Association (AAA) and advise them of the complaint. The AAA will then assist the parties in the selection of one (small cases) or three (larger cases) arbitrators to hear both sides of the dispute.

Another method for selecting the arbitrators is to have each party select "their" arbitrator and then these two arbitrators would in turn select the "neutral" arbitrator.

No matter how the arbitration panel is chosen, each party will have the opportunity to present their side of the issue and then the panel will render a decision which is binding upon the parties (if specified in the contract).

Litigation—When two parties to a contract have a dispute which is silent on arbitration and cannot agree on any of the above methods to resolve the issue, the remaining option is to enter the judicial system. Depending on the legal residency of the parties and the amount of the claim, the dispute would either be heard in a local court or a federal court. The trial will be held in front of a jury unless both parties agree to waive their right to a jury or specify a Judicial Hearing Officer.

Bibliography

1. The Engineering Management Certification International Website, www.engineeringcertification.org
2. American Society of Mechanical Engineers (ASME) website for the Professional Practices Curriculum's (PPC) Project Management Series, www.professionalpractice.asme.org
3. *Construction Executive Magazine*, November 2003, page 9.
4. Advanced Engineering Economics, Park, John Wiley and Sons, 1990 Chap 12.
5. The Business Roundtable, Report A-7, Appendix B, Contractual Arrangements, Construction Industry Cost Effectiveness Project Report
6. ASME website for the PPC's Safety and Risk Assessment Module, www.professionalpractice.asme.org
7. Association for the Advancement of Cost Engineering International (AACEI) Recommended Practice for Cost Estimate Classification, AACE 17R-97 and AACE 18R-97 for the process industries.
8. Construction Industry Institute, Special Publication 17-1, in search of Partnering Excellence, Table 2.1, July 1991.
9. Dr. Edwards Deming's 14 points.
10. Construction Accounting Manual, Risk Memos, Figure 11A-6.
11. Glossary (excerpted from R.S. Means and AACEI Glossaries).
12. "The Deadly Dozen," *Journal of Construction Accounting and Taxation*, Fall 1991.
13. Construction Industry Institutes "Productivity Measurement: An Introduction," Publication 2–3, 1990.

About the Author

Ronald Saporita, P. E.

Mr. Saporita has over 40 years worldwide experience in planning, engineering, construction and operation of industrial facilities, as well as the support and infrastructure required. His experience includes commercializing various products and processes, management of commissioning of chemical and petrochemical facilities, and project management and corporate oversight of capital process and manufacturing improvement projects; including specialty chemicals, catalysts, desiccants, multilayer films for food packaging, container sealants, construction products, medical devices, transportation, and various building functions (schools, hospitals, research, office and commercial). He was also the corporate team leader for the safety risk reviews and investigation of chemical plant incidents.

His most comprehensive assignment was as corporate team leader for the evaluation and selection of a grass roots Asian site, including negotiating for the land and government incentives, permitting, long-range (20 year) planning, and obtaining corporate budgetary approval, selection of contractors for engineering and construction, and project management for the design, construction and commissioning of the first phase.

Most recently he was Vice President of Wilson Management Associates, Inc. (WMA) in Glen Head, New York, responsible for assisting clients in the analysis and reduction of various types of risks, as well as resolution of disputes for design and construction projects. Typical disputes involve cost and schedule overruns.

Prior to joining WMA, he was Vice President of the NYC School Construction Authority, responsible for project management and support services such as environmental, permitting, scheduling, cost control, quality and safety. Prior to that, he was the Director of International Engineering for W. R. Grace & Company, and Manager of the Technical Services Division of Stone & Webster Engineering Corporation, and a Process Equipment Design Engineer at the M.W. Kellogg Company.

He has a B. E. and M. E. in Mechanical Engineering from the Cooper Union, is a Fellow of the American Society of Mechanical Engineers, and is a licensed Professional Engineer in several states. Mr. Saporita is a Member of ASME's Board of Professional Development, past Chairman of the Safety and Risk Analysis Division, former team leader and contributor to ASME's Professional Practices Curriculum Initiative, and a founding member of the Engineering Management Certification International Initiative. He was the recipient of ASME's Dedicated Service Award in 2002.

Glossary of Construction Terms[11]

activity An operation or process consuming time and possibly resources. An activity is an element of work that must be performed in order to complete a project. An activity consumes time, and may have resources associated with it. Activities must be measurable and controllable. An activity may include one or more tasks.

activity duration The length of time from start to finish of an activity, estimated or actual, in working or calendar time units.

addendum, addenda (pl.) Alteration or clarification of the plans or specifications provided to the bidder by the owner or by the owner's representative prior to bid time. An addendum becomes part of the contract documents when the contract is executed.

advertisement for bids The published, public notice soliciting bids for construction. Usually, it is required by law that the advertisement be published in newspapers of general circulation in the area, when public funds are to be used for construction.

agent A person authorized by another person to act on the first person's behalf. An architect is frequently the owner's agent.

agreement (1) A meeting of the minds. (2) A promise to perform, between signatories of a document. (3) In construction, the specific documents setting forth the terms of the contracts between architect, owner, engineer, construction manager, contractor, and others.

allowance Additional resources included in estimates to cover the cost of known but undefined requirements for an individual activity, work item, account or subaccount.

alternate bid An amount stated in a bid which can be added or deducted by an owner if the defined changes are made to the plans or specifications of the base bid.

arbitration A method of settling disputes between parties of a contract by presenting information to recognized authorities. Parties agree in advance to binding arbitration of disputes, either as a clause in the contract or at the occurrence of a dispute.

base bid The amount of money stated in the bid as the sum for which bidder offers to perform the work described in the bidding documents, prior to adjustments for alternate bids that have been submitted.

baseline In project control, the reference plans in which cost, schedule, scope and other project performance criteria are documented and against which performance measures are assessed and changes noted.

basic design package Further development of Conceptual Engineering which enables one to be able to solicit independent confirmation of the project costs and timing (when construction or manufacture is involved).

bid A complete and properly signed proposal to do the work or designated portion thereof for the amount or amounts stipulated therein. A bid is submitted in accordance with the bidding documents.

- ***bid bond*** A form of bid security executed by the bidder or principal and by a surety to guarantee that the bidder will enter into a contract within a specified time and furnish any required performance bond, and labor and material payment bond.
- ***bid date*** The date established by the owner or the architect for receipt of bids.
- ***bid form*** A form, furnished to the bidder, on which to submit his bid.
- ***bid opening*** The opening and tabulation of bids submitted within the prescribed bid time and in conformity with the prescribed procedures.

bidding documents The advertisement or invitation to bid, instructions to bidders, the bid form, other sample bidding and contract forms, and the proposed contract documents, including any addenda issued prior to receipt of bids.

bidding period The calendar period beginning at the time of issuance of bidding documents and ending at the prescribed bid time.

bonding company A firm providing a surety bond for work to be performed by a contractor payable to the owner in case of default of the contractor. (*See* guaranty bond)

building code The legal minimum requirements established or adopted by a governmental unit pertaining to the design and construction of buildings.

business agent An official of a trade union who represents the union in negotiations and disputes and checks jobs for compliance with union regulations and union contracts.

change order A written order to a contractor with the necessary signatures to make it a legal document and authorizing a change from the original plans, specifications, or other contract documents, as well as a change in the contract cost or duration. (Zero and negative cost change orders are included.)

clarification of information (postbidding documents) An illustration or description, provided by an architect or engineer, to explain in more detail some area or item on the bid or contract documents, or as part of a job change order or modification.

clerk of the works A representative of the architect or owner who oversees construction, handles administrative matters, and ensures that construction is in accordance with the contract documents.

closed list of bidders A list of contractors that have been approved by the architect and owner as the only ones from whom bid prices will be accepted.

closed specifications Specifications stipulating the use of specific or proprietary products or processes without provision for substitution.

completion date The date established in the contract for completion of all or specified portions of the work. This date may be expressed as a calendar date or as a number of days after the date for commencement of the contract time is issued.

completion list The final list of items of work to be completed or corrected by the contractor. Sometimes called a "punch list."

conceptual engineering Further development of the concept from the Planning Study to arrive at a decision to invest (apply additional resources).

construction documents Drawings and specifications setting forth in detail the requirements for the construction of a project.

contingency An amount added to an estimate to allow for items, conditions, or events for which the state, occurrence, and/or effect is uncertain and that experience shows will likely result in aggregate, in additional costs. Typically estimated using statistical analysis or judgment based on past asset or project experience. Contingency usually excludes: (1) major scope changes such as changes in end product specification, capacities, building sizes, and location of the asset or project, (2) extraordinary events such as major strikes and natural disasters, (3) management reserves, and (4) escalation and currency effects. Some of the items, conditions or events for which the state, occurrence, and/or effect is uncertain include, but are not limited to, planning and estimating errors and omissions, minor price fluctuations (other than general escalation), design developments and changes within the scope, and variations in market and environmental conditions. Contingency is generally included in most estimates, and is expected to be expended.

contingent agreement An agreement, generally between an owner and an architect, in which some portion of the architect's compensation is contingent upon the owner's obtaining funds for the project (such as by successful referendum, sale of bonds, or securing of other financing), or upon some other specially prescribed condition.

contract documents The agreement, addenda (which pertain to the contract document documents) contractor's bid (including documentation accompanying the bid and any postbid documentation submitted prior to the notice of award) when attached as an exhibit to the agreement, the bonds, the general conditions, the supplementary conditions, the specifications and the drawings as the same are more specifically identified in the agreement, together with all amendments, modification and supplements issued pursuant to the general conditions on or after the effective date of the agreement.

critical path One or more sequences of activities with the least amount of total float activities running from the start event to the finish event in the schedule. It is the longest time path through the schedule.

damages Usually, per diem amounts specified in a contract and payable only when incurred loss can be proved to have resulted from a contractor's delays or breach of contract.

date of agreement The date shown on the face of an agreement, or the date the agreement is signed. It is usually the date of the award.

date of commencement of the work The date established in the notice to proceed or, in the absence of such notice, the date of the agreement or such other date as may be established therein or by the parties thereto.

date of substantial completion The date certified by the architect when the work or a designated portion thereof is sufficiently complete, in accordance with the contract documents, so that the owner may occupy the work or designated portion thereof for the use for which it is intended.

deliverables A performance standard for each subunit or task. Examples are drawings or specifications, or finished products.

early finish The earliest time at which an activity can be completed; equal to the early start of the activity plus its remaining duration.

early start The earliest time any activity may begin as logically constrained by the network for a specific work schedule.

extended coverage insurance Work-site insurance (included in property insurance) against loss or damage caused by wind, hail, riot, civil commotion, aircraft, land vehicles, smoke, and explosion (except steam boiler explosion).

extra work Any work, desired or performed, but not included in the original contract.

feasibility study A thorough study of a proposed construction project to evaluate its economic, financial, technical, functional, environmental, and cultural advisability.

final acceptance The formal acceptance of a contractor's completed construction project by the owner, upon notification from an architect that the job fulfills the contract requirements. Final acceptance is often accompanied by a final payment agreed upon in the contract.

final completion A term denoting that the work has been completed in accordance with the terms and conditions of the Contract Documents.

final payment The payment an owner awards to the contractor upon receipt of the final certificate for payment from the architect. Final payment usually covers the whole unpaid balance agreed to in the contract sum, plus or minus amounts altered by change orders.

force account A term for any work ordered on a construction project without an earlier agreement on its lump sum or unit price cost and performed with the understanding that the contractor will bill the owner according to the cost of labor, materials, equipment, insurance, and taxes plus a certain percentage for overhead and profit.

front end loaded bid A contractor's bid based on increased costs for tasks to be performed early and decreased costs for later tasks. The front end bid is used to improve a contractor's cash flow.

general conditions The portion of the contract documents in which the rights, responsibilities, and relationships of the involved parties are itemized.

general construction contractor In a multiple prime contract delivery method in New York, relates to the work exclusive of mechanical, electrical and plumbing trades.

general contractor For an inclusive construction project, the primary contractor who oversees and is responsible for all the work performed on the site, and to whom any subcontractors on the same job are responsible.

general foreman The general contractor's on-site representative, often referred to as the "superintendent" on large construction projects. It is the responsibility of the general foreman to coordinate the work of various trades and to oversee all labor performed at the site.

guarantee A legally enforceable assurance of quality or performance of a product or work, or of the duration of satisfactory performance. Also called "guaranty" and/or "warranty."

guaranteed maximum price The maximum amount above which an owner and contractor agree that cost for work performed (as calculated on the basis of labor, materials, overhead, and profit) will not escalate.

guaranty bond Each of the four following bonds are types of guaranty bonds: (1) bid bond, (2) labor and material payment bond, (3) performance bond, (4) surety bond and (5) maintenance bond.

hold harmless A clause of indemnification by which an insurance carrier agrees to assume his client's contractual obligation and to assume responsibility in certain situations which otherwise might be the obligation of the other party to the contract.

insurance A contractual obligation by which one person or entity agrees to secure another against loss or damage from specified liabilities for premiums paid.

- ***professional liability*** Insurance coverage for the insured professional's legal liability for claims for damages sustained by others allegedly as a result of negligent acts, errors, or omissions in the performance of professional services.
- ***workers' compensation*** Insurance covering the liability of an employer to employees for compensation and other benefits required by worker's compensation laws with respect to injury, sickness, disease, or death arising from their employment. Also still known in some jurisdictions as "workmen's compensation insurance."
- **employer's liability insurance** Insurance that protects an employer from his employees' claims for damages resulting from sickness or injury sustained during their course of work and based on negligence of common law rather than on liability under workmen's compensation.

- **comprehensive general liability insurance** A broad form of liability insurance covering claims for bodily injury and property damage which combines under one policy coverage for all liability exposures (except those specifically excluded) on a blanket basis and automatically covers new and unknown hazards that may develop. Comprehensive general liability insurance automatically includes contractual liability coverage for certain types of contracts. Products liability, completed operations liability and broader contractual liability coverage are available on an optional basis. This policy may also be written to include automobile liability.

invitation to bid A portion of the bidding documents soliciting bids for a construction project.

labor and material payment bond A contractor's bond in which a surety guarantees to the owner that the contractor will pay for labor and materials used in the performance of the contract. The claimants under the bond are defined as those having direct contracts with the contractor or any subcontractor. A labor and material payment bond is sometimes referred to as a payment bond.

latent defect A defect in materials or equipment that could not have been discovered under reasonably careful observation. A latent defect is distinguished from a patent defect, which may be discovered by reasonable observation.

latest start date The latest time at which an activity can start without lengthening the project.

letter of intent A letter signifying an intention to enter into a formal agreement, usually setting forth the general terms of the agreement.

licensed contractor A person or entity certified by a governmental authority, where required by law, to engage in construction contracting.

lowest responsive bid The lowest bid that is responsive to and complies with the requirements of the bidding documents.

maintenance bond A contractor's bond in which a surety guarantees to the owner that defects of workmanship and materials will be rectified for a given period of time. A one year bond is commonly included in the performance bond.

maintenance period The period after completion of a contract during which a contractor is obligated to repair any defects in workmanship and materials that may become evident.

mechanic's lien A lien on real property, created by statute in all states, in favor of persons supplying labor or materials for a building or structure for the value of labor or materials supplied by them. In some jurisdictions a mechanic's lien also exists for the value of professional services. Clear title to the property cannot be obtained until the claim for the labor, materials or professional services is settled.

milestones When depicting a project schedule, it sometimes becomes necessary to introduce a milestone in order to preserve the accuracy of the representation. A milestone is a fictitious activity of null duration. It has a cost of zero. However, it does have a reality in that it may not commence before any of its precedents are completed and none of its successors may begin until it has been completed. It is usually reserved to demonstrate a significant event. Examples are "Ready for Testing" or "Receipt of Material."

negligence Failure to exercise due care under the circumstances. Legal liability for the consequences of an act or omission frequently depends upon whether or not there has been negligence.

network A logic diagram of a project consisting of the activities and events that must be accomplished to reach the objectives, showing their required sequence of accomplishments and interdependencies.

notice to bidders A notice contained in the bidding documents informing prospective bidders of the opportunity to submit bids on a project and setting forth the procedures for doing so.

notice to proceed Written communication issued by the owner to the contractor authorizing the work to proceed and establishing the date of commencement of the work.

open bidding A bidding procedure wherein bids or tenders are submitted by and received from all interested contractors, rather than from a select list of bidders privately invited to compete.

partial occupancy Occupancy by the owner of a portion of a project prior to completion of the balance of work.

penal sum The amount named on a contract or bond as the penalty to be paid by a signatory thereto in the event that the contractual obligations are not performed.

performance bond (1) A guarantee that a contractor will perform a job according to the terms of the contract, or the bond will be forfeited. (2) A bond of the contractor in which a surety guarantees to the owner that the work will be performed in accordance with the contract documents. Except where prohibited by statute, the performance bond is frequently combined with the labor and material payment bond.

physical progress A measurement of actual items not related to the cost. For example, the number of drawings issued that can be related to the total expected.

planning study The initial technical and financial evaluation of a concept to determine if additional investment in resources is warranted.

predecessor An activity that immediately precedes another activity.

principal-in-charge The architect or engineer in a professional practice firm charged with the responsibility for the firm's services in connection with a given project.

project manager A term frequently used to identify the individual designated by the principal-in-charge to manage the firm's services related to a given project. Normally, these services include administrative responsibilities as well as technical responsibilities.

quality Conformance to establish requirements (not a degree of goodness).

quality control A system of procedures and standards by which a constructor, product manufacturer, material processor, or the like, monitors the properties of the finished work.

record drawings Construction drawings revised to show significant changes made during the construction process, usually based on marked-up prints, drawings, and other data furnished by the contractor to the architect. Completed record drawings are referred to as as-built drawings.

resident engineer An engineer employed by the owner to represent the owner's interests at the project site during the construction phase. The term is frequently used on projects in which a governmental agency is involved.

resources Capital or human resources.

retention Usually refers to a percent of contract value (usually 5 or 10%) retained by the purchaser until work is finished and fully accepted.

schedule of values A statement furnished by the contractor to the architect reflecting the portions of the contract sum allocated to the various portions of the work and used as the basis for reviewing the contractor's applications for payment.

scope The sum of all that is to be or has been invested in and delivered by the performance of an activity or project. In project planning, the scope is usually documented (i.e. the scope document), but it may be verbally or otherwise communicated and relied upon. Generally limited to that which is agreed to be the stakeholders in an activity or project (i.e., if not agreed to, it is "out of scope"). In contracting and procurement practice it includes all that an enterprise is contractually committed to perform or deliver.

sealed bid A bid, based on contract documents, that is submitted sealed for opening at a designated time and place.

shop drawings Drawings, diagrams, schedules and other data specially prepared for the work by the contractor or any subcontractor, manufacturer, supplier, or distributor to illustrate some portion of the work.

special conditions A section of the conditions of the contract, other than general conditions and supplementary conditions, which may be prepared to describe conditions unique to a particular project.

specifications A part of the contract documents contained in the project manual consisting of written requirements for materials, equipment, construction systems, standards, and workmanship. Under the Uniform Construction Index, the specifications comprise 16 divisions.

subsurface investigation The soil boring and sampling program, together with the associated laboratory tests, necessary to establish subsurface profiles and the relative strengths, compressibility, and other characteristics of the various strata encountered within depths likely to have an influence on the design of the project. Sometimes called "geotechnical investigation."

successor An activity that immediately succeeds another activity.

superintendent The contractor's representative at the site who is responsible for continuous field supervision, coordination, completion of the work, and, unless another person is designated in writing by the contractor to the owner and the architect, for the prevention of accidents.

surety A person or entity who promises in writing to make good the debt or default of another.

surety bond A legal instrument under which one party agrees to answer to another party for the debt, default or failure to perform of a third party.

survey (1) To do boundary, topographic and/or utility mapping of the site. (2) To measure an existing building. (3) To analyze a building for the use of space. (4) To determine the owner's requirements for a project. (5) To investigate and report on required data for a project.

test pit An excavation made to examine the subsurface conditions on a potential construction site. Also, a pit excavated to inspect the condition of existing foundations.

time Term defined in reference to a construction contract as time limits or periods stated in the contract. A provision in a construction contract that "time is of the essence of the contract" signifies that the parties consider punctual performance within the time limits or periods in the contract to be a vital part of the performance. Failure to perform on time is a breach for which the injured party is entitled to damages in the amount of loss sustained.

total float The amount of time (in work units) that an activity may be delayed from its early start without delaying the project finish date. Total float is equal to the late finish minus the early finish or the late start minus the early start of the activity.

unbalanced bid A contractor's unit price bid based on an expectation of the increase in quantities estimated for the more high-cost items. Note: This is a source (often) of disputes.

value engineering A process of reviewing plans and specifications for the purpose of reducing the final cost without changing the intended utility or overall appearance.

Work Breakdown Structure A division of work which organizes, defines and displays all of the work to be performed in accomplishing the project objectives.

zoning The reservation of certain specified areas within a community or city for buildings and structures, or of use of land, for certain purposes with other limitations, such as height, lot coverage, and other stipulated requirements.

Appendix 1

Goals Matrix

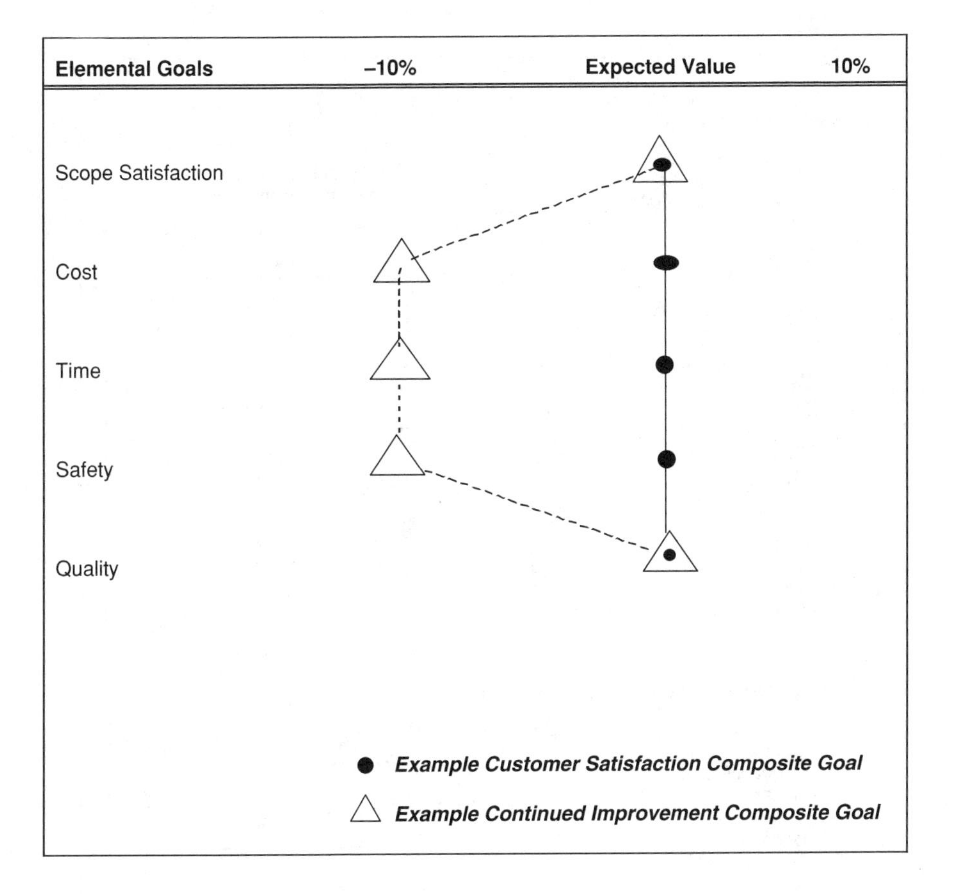

Appendix 2

Project Risk Management Process

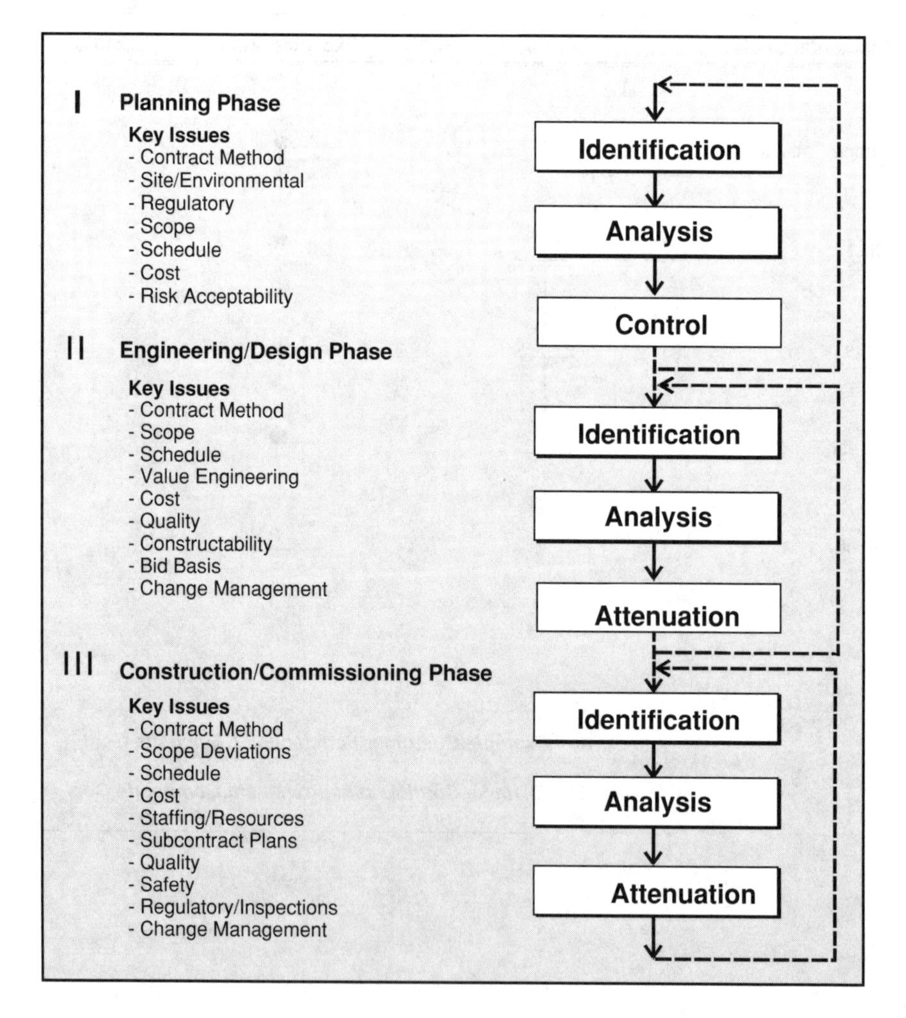

Appendix 3

Record Keeping:
Important Construction Documents

(Contracts, project baseline schedule and budgets are discussed separately in the text.)

1. **DRAWING REVISIONS:** During design and construction, it is essential that drawing revisions be bubbled, or identified each time a change is made. As-built drawings should be the result of all design revisions and acceptable field deviations.
2. **PROJECT SPECIFICATIONS:** The detailed requirements for systems and materials that are not on the contract drawings, are incorporated in the specifications. Revisions or disputed interpretations must be maintained as a mark-up or insert to the contract set.
3. **REQUISITIONS:** This document is a historical record of work completed throughout the project. It should reasonably reflect the weighted percentages of each line item on the work breakdown structure or schedule of values. This applies to the prime contract and all subcontracts.
4. **CHANGE ORDERS:** There is never too much information when documenting a change order. Every change order should reference contract documents and indicate all impacts (reduction/increase in scope, time and cost), as well as the reason (field condition, scope change, drawing error, material substitution, etc.). A time extension explanation should refer to the CPM logic.
5. **CORRESPONDENCE:** The main purpose of the correspondence is to timely address one specific issue as a discussion or notice. Support the position as stated in the letter by identifying relevant documents and impacts. This should facilitate communication based on the significance of the issue discussed and preserve rights for disputed items.
   Particular attention should be paid to timing of notice letters that preserve rights to resolution of disputed costs and/or extensions of time.
6. **CLARIFICATION MEMOS (Request For Information/RFI):** Throughout the course of a project a clarification is often needed

from the owner or designer. The submitted requests for clarification, along with the response, should be in writing and contain specific references to the questionable drawing or specification.

7. **STATUS LOGS (for shop drawing submittal and approval):** A system of logs, either manual or computerized, to control the flow of the various documents throughout the project, presents a good view of the timeliness and interaction of the drawing approval process, RFI process and change order process. Any RFI or change order impact should be cross-referenced on the log.
8. **TRANSMITTAL LETTERS/LOG:** Anything sent to other parties (i.e., drawings, change orders, pay requisitions, etc.) should be sent with a short transmittal letter or form, indicating who it is being sent to, why it is being sent, and the day it is being sent. If a response is required, it should be indicated on the transmittal form, including a reasonable expected timing of the response.
9. **DAILY PROGRESS REPORTS:** "Standard" daily reports are not designed for a particular project. The daily progress report must meet the particular needs of the project and be as specific as possible when citing work and the associated locations. It will contain information that will support other documents (requisitions, schedules, etc.). This document should contain the most descriptive information for evaluating impacts.
10. **MINUTES OF MEETINGS:** An agenda should be published for every meeting and minutes should be published shortly after the meeting is complete. These minutes should reflect discussion of all impacts, including disruptions and delays. If they contain errors, written clarification should be provided within time limits agreed to. Because the daily progress reports should support the discussion in the minutes, the minutes are a starting point when reviewing impacts after the fact.
11. **STATUS REPORTS:** Periodic status reports, on a weekly or monthly basis, should contain a discussion of significant issues. The discussion should be a summary of the typical detailed issues contained in meeting minutes and include copies of project summary level schedules and cost reports. Any discussion of impacts should, at least, refer to a meeting minute discussion.
12. **PHOTOGRAPHS:** There is no better way to document what is happening on a construction project than with the use of photographs or video equipment. The photos must contain a description and be dated. It is often useful to periodically take photos from fixed locations.

Appendix 4

Types of Contracts

Options	Advantages	Disadvantages	Contractor Relationships	Comments
I. COST REIMBURSABLE (AND CONVERTIBLE)				
A. Cost Reim. + % Fee	• Maximum contracting flexibility • Can fast track easier • Best for development/ changing scope • Easier to insert proprietary needs	• Required financial controls • Requires highest owner staffing • Many interfaces to manage • Limited incentive by contractor to contain cost • Owner assumes all risks	• Collaborative • Owner can substantially influence contractor decisions; assumes more responsibility for results	• Use with poorly defined scope • Use with known, trusted contractor resources
B. Cost Reim. + Fixed Fee	• Rework provides no profit	• Limited contractor profits may invite less staffing (quality or quantity)	• As above	• As above

C. Cost Reim. + Incentive Fee	• As above, but relates fee to performance and provides contractor with incentives to control cost and schedule • Allows alignment of owner and contractor goals	• Difficult to establish incentive goals that differentiate outstanding performance from good/normal performance • Scope changes impact on incentives requires management and control	• Contractor will "push back" on items affecting their fee	• Can results be achieved w/o incentives? • Good negotiating skills required • Incentive must relate to one key project objective (never to both budget and schedule, for example)
D. Cost Reim. W/Guaranteed Maximum	• Knowing top side cost risk in advance	• Required scope to be more than 80% defined • Changes must be documented and cost negotiated • Maintaining profit margin will be prime objective of contractor	• Adversarial on changes	• See discussion in text

(*Continued*)

Options	Advantages	Disadvantages	Contractor Relationships	Comments
E. Convertible Contract (start cost reim., then convert to fixed price)	• Reduces owner's cost risk to definition phase. Allows opportunity to negotiate or bid subsequent phases • Provides time to reduce uncertainties	• Requires skillful owner negotiation	• Collaborative	• Transion negotiations may impact schedule.
F. Unit Price w/ Adjustments	• Use when scope is clear except for quantities • Cost risk reduced to quantity impact of material price changes	• Scope & quality must be controlled • Owner takes price risk for quantity increases	• Collaborative	• Used mainly for heavy construction.

II. FIXED PRICE				
A. Lump Sum	• Contractor assumes cost risk • Allows competitive bidding on total scope • Incentive for contractor to furnish best resources • Can test unknown contractors with minimum risk	• Quality and schedule risk increased • Complete definition required up front • Bidding process will lengthen total schedule • Change negotiations may require additional owner resources	• Contractor's profit at risk	• Good for known technology projects • Difficult to define sufficiently for development activities • Early/preliminary design can be done reimbursable to protect schedule
B. Multiple Lump Sum	• More competitive pricing	• Contract interfaces must be clear • Project completion dependent on slowest contractor • Good, complete engineering package required	• Owner has overall project accountability	• Used in a competitive labor market

(Continued)

Options	Advantages	Disadvantages	Contractor Relationships	Comments
C. Fixed Price w/ Incentives	• Further reduce cost where scope is susceptible to lower cost option/ methods	• Requires changing contract documents and negotiating new price	• Contractor will push back on items negatively impacting incentive amount	• Used where the owner has experience with similar projects.
D. Fixed Price w/Escalation	• Contractor does not have to provide large contingencies to out guess inflation	• Owner carries inflation risk	• Lowers contractor risk	• Used where inflation is high or project start uncertain
E. Design/Build	• Single interface management. • Single contractor accountability • Able to buy contractor's special knowledge	• More technical reliance on contractor • Quality may be lower than expected • Must know construction site labor situation • Change orders may require more time to resolve	• Contractor responsible for errors in design • Designer/Contractor team experience required	• Faster completion depends on previous project and team experience • Excellent for repeat projects • Owner can purchase special equipment

F. Design/Build/ Operate (Turnkey)	• Contractor assumes all cost and performance risk • Provides opportunity for new state-of-the-art technology • Less owner staffing	• Must accept contractor's technical solution • Difficult to prespecify all performance criteria (operating cost, maintenance, clean design) • No timely recourse if contractor fails	• Owner must trust contractor for end result • Mfg. must accept installation "as is" if performing	• Recognize finished installation might not meet all expectations • "Industry standard" result is normal • Contractor should procure everything • Performance guarantee by contractor has limitations

Appendix 5

Cost Impact of Controllable Risks

	COST ACCOUNTABILITY		COST IMPACT TO OWNER	
	FIXED PRICE	REIM-BURSABLE	FIXED PRICE	REIM-BURSABLE
CONTROLLABLE RISKS				
1. Labor Productivity			**Low**	**Medium**
a. Management of work force	Contractor	Owner		
b. Timing and quality of Engineering data and Equipment	Owner/ Engineer	Owner/ Engineer		
c. Quality assurance	Owner	Owner		
d. Quality control	Owner	Owner		
2. Scope			**High**	**Low**
a. Initial scope definition	Owner/ Engineer	Owner/ Engineer		
b. Changes in scope	Owner/ Engineer	Owner/ Engineer		
3. Indirect Costs			**Low**	**High**
a. Staff	Contractor	Owner		
b. Consumables	Contractor	Owner		
c. Support Crafts	Contractor	Owner		
d. Materials management	Contractor	Owner		

	COST ACCOUNTABILITY		COST IMPACT TO OWNER	
	FIXED PRICE	REIM-BURSABLE	FIXED PRICE	REIM-BURSABLE
4. Quality Construction			**Medium**	**Medium**
a. Complexity of design	Owner/ Engineer	Owner/ Engineer		
b. Completeness of engineering Drawings	Owner/ Engineer	Owner/ Engineer		
c. Construction procedures and Methods	Contractor	Owner/ Engineer		
d. Construction schedule	Owner/ Engineer	Owner/ Engineer		
e. Experience of craftsmen	Contractor	Owner		
f. Training of craftsmen	Contractor	Owner		
g. Supervisory personnel	Contractor	Owner		
h. Construction equipment and tools	Contractor	Owner		
i. Quality control procedures	Contractor	Owner		
5. Safety			**Medium**	**Medium**
a. Training	Contractor	Owner		
b. Contractor's minimum standards	Contractor	Owner		
c. Owner's mandatory standards	Owner	Owner		
d. Regulatory standards (OSHA, etc)	Contractor	Contractor		
e. Industrial hygiene	Shared	Shared		

(Continued)

	COST ACCOUNTABILITY		COST IMPACT TO OWNER	
	FIXED PRICE	REIM-BURSABLE	FIXED PRICE	REIM-BURSABLE
6. Schedule			**High**	**High**
a. Manufacturer's promised deliveries	Contractor	Owner		
b. Owner-supplied material	Contractor	Owner		
c. Contractor-supplied	Contractor	Owner		
d. Manpower resource	Contractor	Owner		
e. Manpower productivity	Contractor	Owner		
f. Scheduling techniques	Contractor	Owner		
g. Schedule duration	Owner	Owner		
h. Extended overtime or shift work	Owner	Owner		
7. Labor relations			**Low**	**Low**
a. Jurisdictional disputes	Contractor	Owner		
b. Illegal strikes and walk-offs	Contractor	Owner		
c. Contract expiration strikes	Contractor	Owner		
d. Jurisdictional disputes between Contractors	Contractor	Owner		

	COST ACCOUNTABILITY		COST IMPACT TO OWNER	
	FIXED PRICE	REIMBURSABLE	FIXED PRICE	REIMBURSABLE
8. Project management			Low	Low
a. Adequate design drawings	Owner/ Engineer	Owner/ Engineer		
b. Timely procurement and delivery of Materials and Equipment	Owner/ Engineer	Owner/ Engineer		
c. Limitation of number of changes And revisions to drawings and Specifications	Owner/ Engineer	Owner/ Engineer		
d. Quality of fabrication of materials and equipment	Owner/ Engineer	Owner/ Engineer		

The Business Roundtable, Report A-7, Appendix B, Contractual Arrangements, a Construction Industry Cost Effectiveness Project Report.

Appendix 6

Project Work Flow

Phases

Planning Study

Typically used for initial budgeting.

Requires basic understanding of project (process and location experience).

May require up to 5% of total design expenditure.

Key Activities

1. Project definition (customer need)
2. Site requirements data
3. Process Design
4. Environmental, Health and Safety Review
5. Hazards/risk philosophy
6. Process/manufacturing flow diagrams—Major Systems
7. Raw materials/utility requirements
8. Site and facility layout
9. Materials of construction
10. Equipment list—Major Equipment
11. Specifications of major equipment
12. Zoning, permitting and licensing requirements
13. Schedule
14. Budget with 25–40% allowance
15. Resource identification
16. Final evaluation of major objectives and market considerations
17. Program Finalization (based on objectives)

Phases	**Key Activities**
Conceptual Design Typically used for formal position paper to support funding where process and location are familiar. May require up to 10% of total design expenditure.	1. Continued development of the above 2. Instrumentation and control philosophy 3. Establish building HVAC requirements 4. Process hazards analysis 5. Equipment List—Complete 6. Process/Manufacturing Flow Diagrams—All Systems 7. Outside pricing of major equipment 8. Schedule 9. Budget with 15–25% allowance
Basic Design Ideally used for funding approval (if schedule permits). May require up to 20% of total design expenditure and construction funding if Design/Build.	1. Continued development of the above 2. Value engineering 3. Spare parts definition 4. Detailed Layout 5. Complete Material Summary 6. Schedule 7. Budget with 10–15% allowance 8. Broad Constructability Review 9. Confirmation with Program and Objectives 10. Design/Build or detailed design/bid documents

Phases	Key Activities
Detailed Design and Construction Planning Typically begins after funding approval. May require up to 95% of the total design expenditure and a significant amount of time.	1. Construction bid documents (if design/bid/build) 2. Procurement of major equipment (if reimbursable design/construction or Owner-supplied) 3. Identify start-up materials 4. Establish field inspection procedures 5. Establish warehousing responsibilities 6. Review flushing and cleaning requirements 7. Final constructability reviews 8. Schedule 9. Budget with 5–10% allowance
↓	
Construction, Procurement and Precommissioning May require specialized commissioning/programming assistance or regulatory approval and the balance of the total design expenditure. Begin turnover from design and construction team to owner.	1. Procurement (if design/bid/build) 2. Establish turnover schedule and responsibilities 3. Prestartup review plan 4. Establish performance test plan 5. Schedule receipt of raw materials/utilities 6. Compile design documentation and vendor data 7. Obtain Certificate of Occupancy or applicable certification of fit for use 8. Obtain warrantees/guarantees/training 9. Schedule 10. Budget with 5% allowance
↓	

Phases	**Key Activities**
Commissioning Complete turnover to owner.	1. Institute change controls 2. Institute mechanical integrity program 3. Final Close-out of Contract(s)

Appendix 7

Cost Tracking/Forecast Summary

WP	Description	Budgeted Cost[1]	Actual Cost[2]	Percent Complete	Earned Value[2]	Cost to Complete	Forecast at Completion	Budget vs. Completion Variance[2]
1	GCs	$ 6,188,831	$ 5,534,207	69.6%	$ 4,305,274	$ 1,883,557	$ 7,417,764	$ (1,228,933)
2	Transmission Pipeline Materials	$ 9,387,526	$ 7,393,749	100.0%	$ 9,387,526	$ —	$ 7,393,749	$ 1,993,777
3	Transmission Pipeline Valves	$ 479,000	$ 471,375	100.0%	$ 479,000	$ —	$ 471,375	$ 7,625
4	Process Equipment and Yard Piping	$ 9,215,134	$ 6,450,588	99.6%	$ 9,181,038	$ 34,096	$ 6,484,684	$ 2,730,450
5	Clear and Grub	$ 459,723	$ 530,165	76.7%	$ 352,454	$ 107,269	$ 637,434	$ (177,711)
6	Sitework	$ 2,091,004	$ 2,314,505	88.5%	$ 1,849,514	$ 241,490	$ 2,555,995	$ (464,991)
7	Sanitary and Temporary Water	$ 283,824	$ 333,746	94.6%	$ 268,541	$ 15,283	$ 349,029	$ (65,205)
8	Water Bearing Structures	$ 4,603,424	$ 4,451,060	100.0%	$ 4,603,424	$ —	$ 4,451,060	$ 152,364
9	Plant Mechanical	$ 3,638,212	$ 4,046,277	98.8%	$ 3,593,826	$ 44,386	$ 4,090,663	$ (452,451)
10	Admin Building	$ 2,374,973	$ 1,646,427	64.8%	$ 1,538,983	$ 835,990	$ 2,482,417	$ (107,444)
11	Buildings	$ 2,785,286	$ 2,553,601	86.4%	$ 2,405,466	$ 379,820	$ 2,933,421	$ (148,135)

(Continued)

WP	Description	Budgeted Cost[1]	Actual Cost[2]	Percent Complete	Earned Value[2]	Cost to Complete	Forecast at Completion	Budget vs. Completion Variance[2]
12	High Voltage Electrical	$ 2,492,100	$ 2,484,100	100.0%	$ 2,492,100	$ —	$ 2,484,100	$ 8,000
13	Elec Substation Foundation	$ 207,500	$ 154,838	100.0%	$ 207,500	$ —	$ 154,838	$ 52,662
14	Medium/Low Voltage Electrical	$ 6,031,112	$ 5,129,526	86.2%	$ 5,201,834	$ 829,278	$ 5,958,804	$ 72,308
15	Instrumentation	$ 585,000	$ 235,297	77.8%	$ 455,130	$ 129,870	$ 365,167	$ 219,833
16	10MG Storage Tank	$ 2,217,000	$ 2,445,685	100.0%	$ 2,217,000	$ —	$ 2,445,685	$ (228,685)
17	Bottling Plant	$ 347,366	$ 5,555	1.6%	$ 5,555	$ 341,811	$ 347,366	$ —
18	Landscaping	$ 887,337	$ 8,370	2.7%	$ 23,958	$ 863,379	$ 871,749	$ 15,588
19	Fencing	$ 105,980	$ 21,196	20.0%	$ 21,196	$ 84,784	$ 105,980	$ —
20A	42″ Route 1	$ 1,368,876	$ 1,458,708	100.0%	$ 1,368,876	$ —	$ 1,458,708	$ (89,832)
20B	42″ Route 2	$ 890,263	$ 1,078,050	100.0%	$ 890,263	$ —	$ 1,078,050	$ (187,787)
20C	42″ Route 3	$ 766,447	$ 980,758	100.0%	$ 766,447	$ —	$ 980,758	$ (214,311)
20D	84″ Route A	$ 4,104,118	$ 2,979,062	88.4%	$ 3,630,059	$ 474,059	$ 3,453,121	$ 650,997
20E	84″ Route B	$ 2,663,236	$ 2,720,352	100.0%	$ 2,663,236	$ —	$ 2,720,352	$ (57,116)
20F	66″ RWPS	$ 1,415,214	$ 2,935,259	100.0%	$ 1,415,214	$ —	$ 2,935,259	$ (1,520,045)
20G	Road Closure	$ 60,000	$ 85,633	100.0%	$ 60,000	$ —	$ 85,633	$ (25,633)
21	Paving, Curbing	$ 1,338,181	$ 736,000	55.0%	$ 736,000	$ 602,181	$ 1,338,181	$ —

22	Raw Water Intake System	$ 4,519,645	$ 4,709,695	55.8%	$ 2,521,198	$ 1,998,447	$ 6,708,142	$ (2,188,497)
23	Cathodic Protection	$ 482,500	$ 392,810	95.4%	$ 460,165	$ 22,335	$ 415,145	$ 67,355
24	Road Crossing and Tunneling	$ 1,015,125	$ 534,409	100.0%	$ 1,015,125	$ —	$ 534,409	$ 480,716
25	QA/QC	$ 542,615	$ 476,732	93.3%	$ 506,441	$ 36,174	$ 512,906	$ 29,709
26	Permitting Activities	$ 275,000	$ 12,375	4.5%	$ 12,375	$ 262,625	$ 275,000	$ —
27	Noise Testing	$ 10,000		0.0%		$ 10,000	$ 10,000	$ —
28	Construction Surveying	$ 85,000	$ 122,597	98.0%	$ 83,300	$ 1,700	$ 124,297	$ (39,297)
29	Chemical Building	$ 517,178	$ 327,420	84.5%	$ 437,015	$ 80,163	$ 407,583	$ 109,595
30	Drilled Piers	$ 352,272	$ 95,133	100.0%	$ 352,272	$ —	$ 95,133	$ 257,139
31	Commissioning	$ 2,269,819	$ 213,808	9.4%	$ 213,808	$ 2,056,011	$ 2,269,819	$ —
32	**Subtotal:**	**$ 77,055,821**	**$ 66,069,068**	**85.3%**	**$ 65,721,113**	**$ 11,334,708**	**$ 77,403,776**	**$ (347,955)**
33	Contingency[3]	$ 3,206,012	$ 347,955	10.9%	$ 347,955	$ 1,486,602	$ 1,719,410	$ 1,486,602
34	**TOTALS:**	**$ 80,261,833**	**$ 66,417,023**	**N/A**	**$ 66,069,068**	**$ 12,821,310**	**$ 79,123,186**	**$ 1,138,647**

1. Includes approved Change Orders.
2. Costs for this period and associated variance omitted for presentation purposes.
3. See Cost Trend Log.

Appendix 8

Cost Trend Log

ID	Description	Forecast Completion	Contingency Forecast ($)	Trend ($)
1	Committed Costs	$ 65,721,113	$ 328,606[A]	N/A
2	Uncommitted Costs	$ 11,334,708	$ 340,041[B]	N/A
3	Subtotal:	N/A	$ 668,647	$ 668,647
4	Risk Management Log Total[C]	$ 470,000	N/A	$ 470,000
5	Current Forecast Subtotal	N/A	N/A	$ (1,138,647)
6	Budget vs. Completion Variance[D]	N/A	N/A	$ (347,955)
7	Contingency Required for Completion	N/A	N/A	$ (1,486,602)
8	Budget Contingency	N/A	N/A	$ 3,206,012
9	Contingency Available for Completion[E]			$ 1,719,410

Abbreviations: N/A, not applicable.

A. Summary of Earned Value Costs at 1/2%—Attachment 7, Line 32.
B. Summary of Cost to Complete at 3%—Attachment 7, Line 32.
C. See Risk Management Log.
D. Budget vs. Completion Variance—Attachment 7, Line 32.
E. Contingency Forecast at Completion—Attachment 8 — Line 7.

Appendix 9

Construction and Commissioning Phase Risk Management Log

January 03					
Risk Issue	**Forecast Amount of Risk ($)**	**Risk Type**	**Risk Control**	**Responsibility**	**Due Date**
1. Construction vibration damage to surrounding homes. Multiple claims	$25,000	Safety	Rock Breaking. Engaged engineer to review subcontractor means and methods. Monitoring in-place.	Current complaints have been passed to Subcontractor's Insurance Companies. Agents assessing claims for Controls Manager.	15 Feb 03
2. Late contract completion	$35,000	Schedule	Subcontractor is delivering progress to recovery plan for main piping installation (onsite).	Project Manager to pursue contract extension of time for added scope.	31 Mar 03
3. Continuing work on verbal Client instruction ahead of written authorization. Funding Source?	$105,000	Cost and Schedule	Factor estimate and first-pass CPM incorporated into forecast to expedite change order.	Controls Manager to continue to manage and update formal change order requests.	First of Every Month

4. Warranty exposure with Mechanical Subcontractor due to delayed electrical contract completion	$30,000	Cost	Pursue owner change order to include warranty costs for revised final completion date.	Procurement Manager to confirm warranties in subcontracts will begin at revised final completion date. Controls Manager to issue formal change order request	1 Feb 03 8 Feb 03
5. Underestimation of quantities that form the basis of the GMP	$135,000	Quality of Design	Early reconcilliation of quantities on all subcontracts.	Controls Manager to manage and update GMP contingency.	First of Every Month
6. Exceed Lump Sum Dewatering Costs	$25,000	Cost	Develop mitigation plan—reduce allowances in subcontracts.	Construction Superintendent to centralize subcontracts responsibility.	15 Feb 03
7. Claim—Changed Subsurface conditions	$75,000	Cost	$200,00 subcontractor claim not worth more than $75,000.	Project Manager to obtain supporting documents.	1 Apr 03
8. Commissioning Failures	$40,000	Cost and Schedule	Maintain stored equipment and thorough cleaning of systems prior to commissioning.	Construction Superintendent to provide idle equipment protection plan and systems cleaning and flushing plan consistant with Commissioning Schedule.	15 Feb 03
TOTAL	$470,000				

Appendix 10

Partnering Practices Comparison

Partnering Relationships	*Traditional Practices*
Mutual Trust forms the basis for strong working relationships.	Suspicion and distrust; each party wary of the motives of actions by the others
Shared Goals and Objectives ensure common direction	Each party's goals and objectives, while similar, geared to what is best for them
Open Communication avoids misdirection and bolsters effective working relationships	Communications structured and guarded
Long-Term Commitment provides the opportunity to attain continuous improvement	Single project contracting
Objective Critique geared to candid assessment of performance	Objectivity limited due to fear of reprisal and lack of continuous improvement opportunity
Access to each other's organization; sharing of resources	Limited access with structured procedures and self-preservation taking priority over total optimization
Total Company Involvement Commitment from CEO to team members	Normally limited to project level personnel

Sharing of business plans and strategies	Sharing limited by lack of trust and different objectives
Absence or minimization of Contract Terms that create an adversarial environment	Routine adversarial relationships for self-protection
Integration of administrative systems and equipment	Duplication and/or translation with attendant costs and delays

Construction Industry Institute Special Publication 17-1, In Search of Partnering Excellence, July 1991.

Appendix 11

Dr. Deming's 14 Points

1. Create constancy of purpose toward improvement of product and service, with the aim to become competitive, stay in business, and provide jobs.
2. Adopt a new philosophy. We are in a new economic age. Management must awaken to the challenge, must learn responsibilities, and take on leadership for change.
3. Cease dependence on inspection to achieve quality. Eliminate the need for inspection on a mass basis by building quality into the product in the first place.
4. End the practice of awarding business on a basis of price tag. Instead, minimize total cost. Move toward a single supplier for any one item on a long-term relationship of loyalty and trust.
5. Improve constantly and forever the system of production and service, to improve quality and productivity, and thus constantly decrease costs.
6. Institute training on the job.
7. Institute leadership. The aim of leadership should be to help people, machines and gadgets to do a better job. Supervision of management is in need of overhaul, as well as supervision of production workers.
8. Drive out fear, so that everyone may work effectively for the company.
9. Break down barriers between departments. People in research, design, sales, and production must work as a team to foresee problems of production and in use that may be encountered with the product or service.
10. Eliminate slogans, exhortations, and targets for the work force that ask for zero defects and new levels of productivity without providing the methods.

11. Eliminate work standards (quotas) on a factory floor. Substitute leadership. Eliminate management by objective. Eliminate management by numbers, numeric goals. Substitute leadership.
12. Remove barriers that rob the hourly worker of his right to pride of workmanship. The responsibility of supervisors must be changed from stressing sheer numbers to quality. Remove barriers that rob people in management and engineering of their right to pride of workmanship. This means abolishment of the annual merit rating system.
13. Institute a vigorous program of education, reeducation and self-improvement.
14. Put everybody in the organization to work to accomplish the transformation. The transformation is everybody's job.

Appendix 12

Project Oversight Checklist

PROJECT REFERENCE DATA:

Title:

Division/Product Line:

Responsible Product Line Manager:

Authorized Amount:

Completion (in Commercial Operation) Date:

Current Currency Exchange Rate:

Make-up of Contingency included, Amount/Percentage (%):

Date:

Review Participants:

NOTE:
These Project Reference Data entries provide minimum information on Key project identities.

The italicized entries on review subjects listed below are indicative of method.

Circumstances of the project specific and the review participants will vary this dialogue on Scope, Schedule, Budget, Contract Administration, and Project Management.

The review process is a Team effort with all input respected to maximize the capital program benefit to the Product Line and Corporation.

1.0 SCOPE:

1.1 Plant capacity/key bottlenecks: *Describe parameters of capacity (i.e., Tons/Year of ______, Operating Service Basis, etc.). Provide basic Process Flow Diagrams, Piping & Instrumentation Diagrams, Line Diagrams or Manufacturing Layout Diagrams.*

1.2 Completeness of Documents: *Percent (%) completion of Process Flow Diagrams, Piping & Instrumentation Diagrams, Equipment*

	Specifications, and Equipment Layouts. Have process and utility material and energy balances been "frozen" and complete? Is the project phased in steps? What is product mix?
1.3 Status of EIA:	*If a consultant prepares an EIA report, then draft document should be reviewed by Corporate Legal, EHS, and technical community prior to issuing to regional environmental authorities. Do not include confidential process information such as pressures, temperatures, reaction times. If no EIA is required, determine documentation of current baseline and projected reasonable impacts of the project on plant area and surroundings.*
1.4 Status of Permits:	*Obtaining building and operating permits' issues plus contingency timing should be allowed for in schedule. Seek experience of other corporate businesses (or similar industrial facilities) in the area.*
1.5 Unproven Materials or Equipment:	*If new technology, any technical risks? Has exact technology been demonstrated elsewhere on a similar commercial scale? Any risk with scale-up of any piece of equipment with proven technology? Do we exactly know what is going to be built—how large (rated capacity) and with what auxiliary facilities? What are turn down ratios of individual pieces of equipment?*

(Continued)

1.0 SCOPE: (Continued)

1.6 Status of HAZOPS/QRA: *Has a HAZOPS/QRA review been accomplished at a stage when P&IDs have been completed and "frozen?" Ensure that mechanism exists for inclusion of correction points reviews into P&IDs and/or other project documents.*

1.7 Plans for Future Expansion: *Future capacity expansion along with support infrastructure, utility, warehouse, and site use of any additional product line manufacturing should be described. Master Planning documentation consistency to be established.*

1.8 Raw Material Supply: *Is there sole or multisuppliers of raw materials? Have specifications been developed? What are plant inventories of materials in planned facilities? Any contingency plans in case of transportation strikes? Does the method of receipt and storage cover best options? Any import duty issues?*

1.9 Product Warehouse: *Have warehouse requirements been reviewed with respect to latest narrow rack configuration available? New narrow aisle forklifts? Any requirements for HVAC related to product quality or normal creature comfort issues? Any need for temporary outside warehouse of inventories to meet "just in time" logistic philosophies? What is the criteria for raw material and finished goods inventory?*

1.10 Engineering Definition:	*What engineering costs, for "internal" and "external" services, (as % of total engineering costs) has been expended? Do not include vendor engineering costs associated with purchased equipment.*
2.0 SCHEDULE:	**Attach Project Schedule**
2.1 Allowance for Permitting:	*Insure adequate time for environmental permit, since this tends to be "long lead" item requiring technical input data on air, water, and solid waste discharges. Building (architectural) and operational permit should be anticipated with adequate planning time.*
2.2 Allowance for New Railroad/ Pier or Other Logistics:	*Consider any upgrades in transportation of products and raw materials from the plant site via new facilities such as rail spur or port expansion. Consider drainage/ containment of spills in all modes of raw material and product movements.*
2.3 Allowance for Corporate Approvals:	*Corporate approvals of design developments, change orders, procurement steps, commitments to agreed scope, and execution plan with time periods should be spelled out in Project Procedures Manual between project personnel and engineering contractor. Does a "typical" one exist? Resident representative will facilitate approval of all technical deliverables.*

(Continued)

2.0 SCHEDULE: (Continued)

2.4 Status of Project Control Schedule:

Review overall project schedule prepared by chosen engineering contractor or project managed subcontractors. Review "float" times, long lead equipment items. What are critical paths? How often will project schedule be updated? Is schedule consistent with any licensed technology or confirmation of any parts of the process design by pilot plant studies or vendor tests?

2.5 Status of Start-up Schedule:

Determine operational turnover understanding from construction. Assure adequate plan for training as equipment is installed and commissioning activity.

2.6 Tie-in with Other Activities Outside Contractor's Scope:

Engineer contractor should be made clear on any matrix split of vendor or corporate supply. For example, hardware supply and software development of a process control system, or any licensed process technology. Any metering stations or logistics consideration for delivery of raw materials and utilities or return of any by-products to suppliers or customers.

2.7 Significance for Investment Schedule:

Any government tax credits for new technology, local employment, taxation deferrals, any associated time statutory limits should be fully considered in project schedule. Are there any supplier or customer product contracts that must be met by new manufacturing plant? Are any construction trades contracts up for negotiation?

3.0 BUDGET:	**Attach Capital Budget**
3.1 Status of Project Control Budget:	*What is the accuracy portrayed in budget? How was budget prepared (factored, supplier bids)? Understand basis for the allowances put in for bulks, contingency, escalation, and the engineering hours.*
3.2 Amount of Foreign Equipment:	*Consider globally sourcing major equipment and currency risks. Consider local sources of refurbished equipment "pots/pans" (small, atmospheric, normal materials vessels).*
3.3 Installed Cost of Piping, Insulation, Electrical and Instrumentation as a Percentage (%) of Fixed Assets (minus building costs):	*These percentages can be calculated and compared to other similar projects. Review allowances built into bulks. Are they reasonable to the contractor and review group's experience? Are there other key ratios?*
3.4 Adequacy/Productivity of Local Labor:	*Construction cost estimates should be prepared by contractor based on local conditions and recent project work in the area. External sources should be used to verify local labor rates and any peak seasonal demands.*
3.5 Expiration of Local/National labor controcts:	*Very large projects may deal with labor agreement impact. Contigency plans on schedule/cost?*
3.6 Start-up Budget Issues:	*Determine consensus understanding for cost associated with operational staff, raw material use, spare parts, off-spec product during commissioning, etc.*

(Continued)

4.0 CONTRACT ADMINISTRATION:

4.1 Contract Form (reimbursable, fixed cost, etc.):	*The contract form for service of engineering consultants, construction, etc. needs to be understood by the project team. Is it?*
4.2 Provision for Change in Contract Form for Construction:	*Staged development of contractor services may offer opportunities for optimizing this project cost.*
4.3 Definition of Completion:	*List of deliverables (for engineering only contracts) or turnover responsibilities for construction contracts should be discussed.*
4.4 Warranty/Guarantee Issues:	*Any licensed technology requiring process performance tests? Any vendor supplied equipment required to meet hydraulic testing? Any testing on analytical and process control instrumentation?*
4.5 Secrecy Issues:	*Engineering contractor and vendor suppliers should have secrecyagreements executed as standard procedure.*
4.6 Estimated Engineering Costs as Percentage (%) of Total Project Cost:	*Add corporate "internal" plus engineering contractor plus any licensed technology engineering divided by total project cost. Compare percentage to other similar volume dollar projects.*

5: PROJECT MANAGEMENT:

Attach Organization Chart/Staffing Plan

5.1 Project Organization Chart:	*Project Manager should have interviewed and be comfortable with all lead engineers and project managers in the proposed contractor's organization for the project.*

	Manning of the project should be reviewed with senior management. Is number of people and time commitments of this resource to review contractor's deliverables sufficient?
5.2 Experience with Similar Projects:	*Is contractor experienced in building plants with similar project characteristics? Are there similar related technologies that contractor has designed? Have we built similar size plants using same technology?*
5.3 Financial Authorization Levels of Project Manager:	*Dollar approval level must be established to project manager by Product Line Management.*
5.4 Cost Trending Procedure:	*Contractor should demonstrate project control procedures in Projects Procedures Manual. Need to clearly communicate to contractor its needs for financial cost reporting to management. Does a "typical" one exist?*
5.5 Schedule Variance Protocol:	*Schedule variance provides a comparison between work scheduled and actual work performed, but does not include actual costs.* *In general, contractor should clearly demonstrate that their project schedule variance method is satisfactory or will be modified to meet our needs.*

(Continued)

5.0 PROJECT MANAGEMENT: (Continued)

5.6 Physical Completion Methodology:

How will physical completion be compared to contractor spending? How will cost trending and scheduling variances be used as deviations between physical completion and contractor costs to show both positive and negative results?

Appendix 13

Risk Memos

I. POLICY. It is the policy of the Company that business ventures be reviewed and approved by the Board of Directors in order to evaluate the financial, technical, legal, or other risks involved. Three weeks' lead time is required on all Risk Memos.

II. SCOPE. The review and approval of the Board of Directors is required on Risk Memos covering proposed commitments, contract changes, and additions, such as the following:

- **A.** Fixed-price proposals.
- **B.** Cost contracts.
- **C.** Domestic joint venture contracts. All potential joint ventures require the approval of the Chief Executive Officer.
- **D.** Foreign investments.
- **E.** Foreign joint ventures.
- **F.** Construction management contracts.

III. ELEMENTS TO BE INCLUDED IN RISK MEMOS

- **A.** Domestic Proposals—Risk Memos must include:
  1. Project information, including identification of key parties to the contract or project with explanation of prior business associations.
  2. Bid information, including all dates, a list of potential bidders, an explanation of the reason for bidding, any special marketing implications, and the total amount to be bid.
  3. Bonding and insurance, including an explanation of the type of coverage required, insurance administration, "wrap-up" policies, and hold-harmless provisions.

4. Insurance information and quotations from the Company's broker of record. Quotes from outside brokers are prohibited.
5. Estimating information, including identification of the originating office and personnel involved; an estimate questionnaire covering plans and specifications; a survey of scope by the company and the subcontractors; owner-furnished or subcontractor-furnished materials; detailed assumptions; and an estimate summary indicating the percentage covered by firm bid.
6. Finance information, including credit rating and reporting of *unbonded subcontractors*, cash requirements, other credit checks, financing, and profitability analysis.
7. Risk qualification, including a tabulation of all areas of potential risk, quantification of that risk (a probability estimate (%), the factored amount, by line), the amount in a contingency reserve included in the cost estimate, an acquisition report that includes a cost and margin analysis, and explanation of any changes or differences between the contract acquired and the information submitted in the Risk Memo, and a bid tabulation for all bids submitted.
8. Contract questionnaire, including an explanation of the type of contract, its compatibility with subcontractor contracts, provisions for changes in time or contract value and any price escalation, a copy of the legal department's appraisal (if any), and any special laws or provisions.
9. An indication as to what portion of the contract is similar to recent work, the availability of company personnel, equipment, and materials, future marketing impact, and the ecological impact, if any.
10. Calculation of the return on assets invested.
11. If the program covered by a proposal is expected to continue for a period in excess of one year, a brief

statement of the planned protection against an abnormal escalation in the economy.

12. Any intent to incorporate production or procurement customer offset agreements, including those obligating other corporate organizations.
13. Comments on any proposed changes in accounting policy, practices, or cost allocations considered in preparation of the proposal or in performance of the work.
14. The amounts of any bid, performance, and/or payment bonds (or letters of credit) that are required and whether or not the Company will be required to indemnify the surety company (or bank).
15. A statement of the net worth and working capital of any third party to be indemnified by the Company and the dollar limit of such indemnity.
16. The source of funds for the project. If the project depends on the customer's obtaining financing, give present status and estimate probability of obtaining adequate funds.
17. If financing offered by the profit center requires a reserve in the proposed costs because a fixed interest rate is being offered, a statement of the amount of the reserve and the size of rate fluctuation or other variance that this reserve will cover.
18. Any arrangement(s) made or contemplated with a sales representative relating to the proposal. If there are none, so state.

B. Foreign Commitments—In addition to the above elements, each Risk Memo that covers plans relating to foreign investments, joint ventures, and licenses must include:

1. A brief explanation of the foreign association being proposed.
2. Identification of the foreign associate, company, or entity (compliance with Foreign Corrupt Practices Act).

3. Terms of the intended agreement.
4. Investment required—cash or valuation of data or other assets.
5. The amount of the sales price or costs that will be in foreign currencies and an estimate of the foreign exchange risks.
6. Provisions for the repatriation of profits, capital, and personnel.
7. Management, technical, and operating personnel obligations required by the parties involved.
8. Any effects on the domestic operations resulting from the transfer of work to the foreign entity.
9. Projection of the sales potential.
10. Evaluation of the political stability of the area involved.
11. Economic risks related to monetary and business fluctuations of the area involved.
12. Any other known association(s) with the same or related foreign associates.
13. United States and foreign government approvals needed or already obtained.
14. Analysis of local tax provisions and the impact on the Company's U.S. tax position.
15. Financing terms, such as letters of credit and advance payments. Review of cash flow forecast.
16. Specific insurance requirements.
17. Plans regarding the site survey.

IV. RESPONSIBILITIES. A Risk Memo is to be prepared and distributed as soon as there has been a decision to pursue the opportunity. This decision to pursue is the result of a bid/no-bid review meeting.

A. The Vice-President/Regional Manager in charge of the project under consideration for a bid is responsible for:

1. Assembling and preparing the Risk Memo.
2. Detailing the risks.

B. The legal department is responsible for:

1. Obtaining review signoffs and the signature of the Office of the Chief Executive.

2. Forwarding the completed Risk Memo to the Board of Directors.

3. Maintaining liaison with the Board of Directors to obtain an expeditious response.

C. Functional representatives are responsible for:

1. Providing the data covering their functional area.

2. Highlighting known or anticipated areas of risk.

V. CAPTIONING OF RISK MEMOS

A. When some of the necessary information is still to be determined, the Risk Memo should be captioned "Advance Notice of Risk Memo."

B. As more definitive data become available, the Risk Memo is to be revised and captioned "Amendment No.—to Risk Memo" or "Final Risk Memo."

C. If a decision is reached to cancel the procurement effort, a "Notice of Cancellation of Risk Memo" is to be issued.

D. Guaranteed Maximum Price (GMP) contracts require specific notification and approval of the Chief Executive Officer at the time at which the GMP value is established. Such notification must be in writing and must be made at least three weeks before the bid is submitted.

VI. RISK MEMO DOCUMENTS FORMAT

A. The attached Risk Memo Format [*omitted*] is to be used.

B. The attached Estimating Questionnaire [*omitted*] is to be used to satisfy Section III A.5.

C. The attached Contract Questionnaire [*omitted*] is to be used to satisfy Section III A.8.

Construction Accounting Manual, Research Institute of America, 1989.

Appendix 14

Oversight Process Diagram

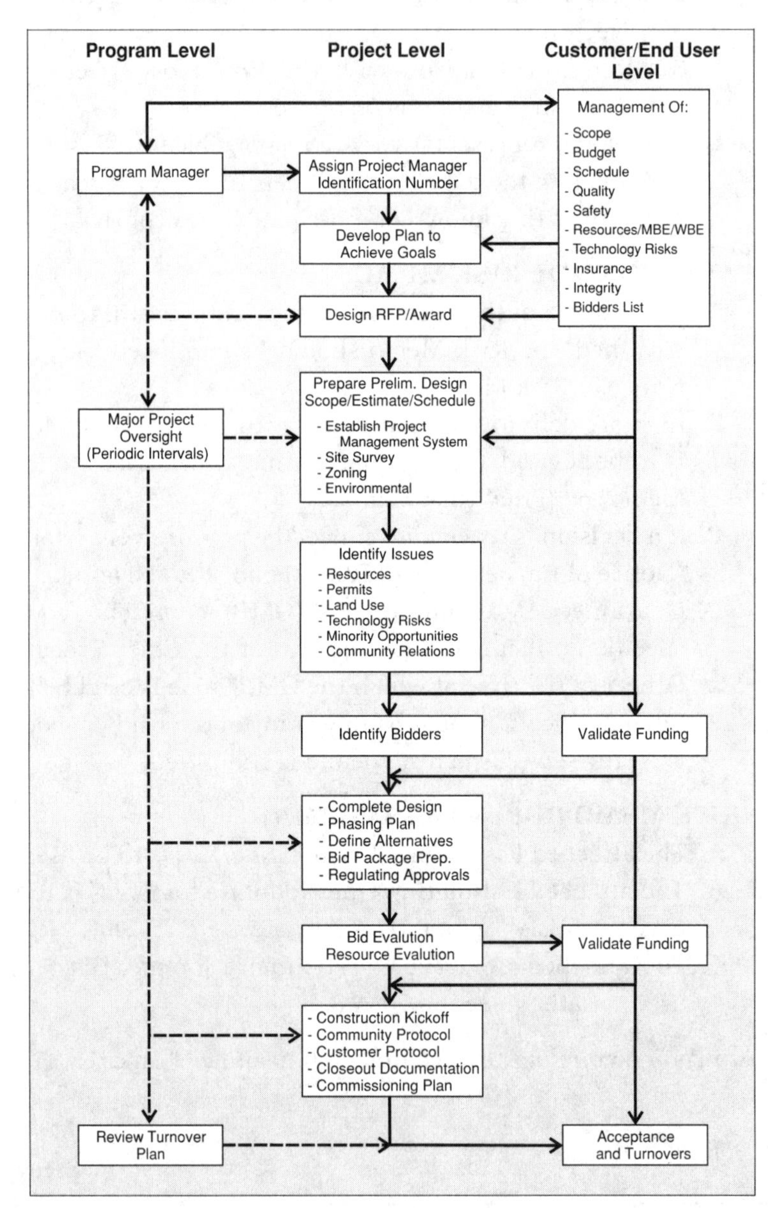

Appendix 15

"Manshul" Delay Damages Example

DAMAGES CALCULATION METHODOLOGY

Contract Value	$754,400		
Completed & Stored	$459,999		
Remaining	$294,401		

Value of work remaining	= $294,401 / 1.15*	= $256,001	(1)
Overhead on (1)	= 0.075 × $256,001	= $19,201	(2)
Profit on (2)	= 0.075 × $19,201	= $ 1,440	
Extended Home Office Damages		= $20,641	

*Adjusted for reasonable overhead and profit (15%) included in the contract price.

Appendix 16

Acceleration Model: Productivity Issues

Acceleration Model: Productivity Issues

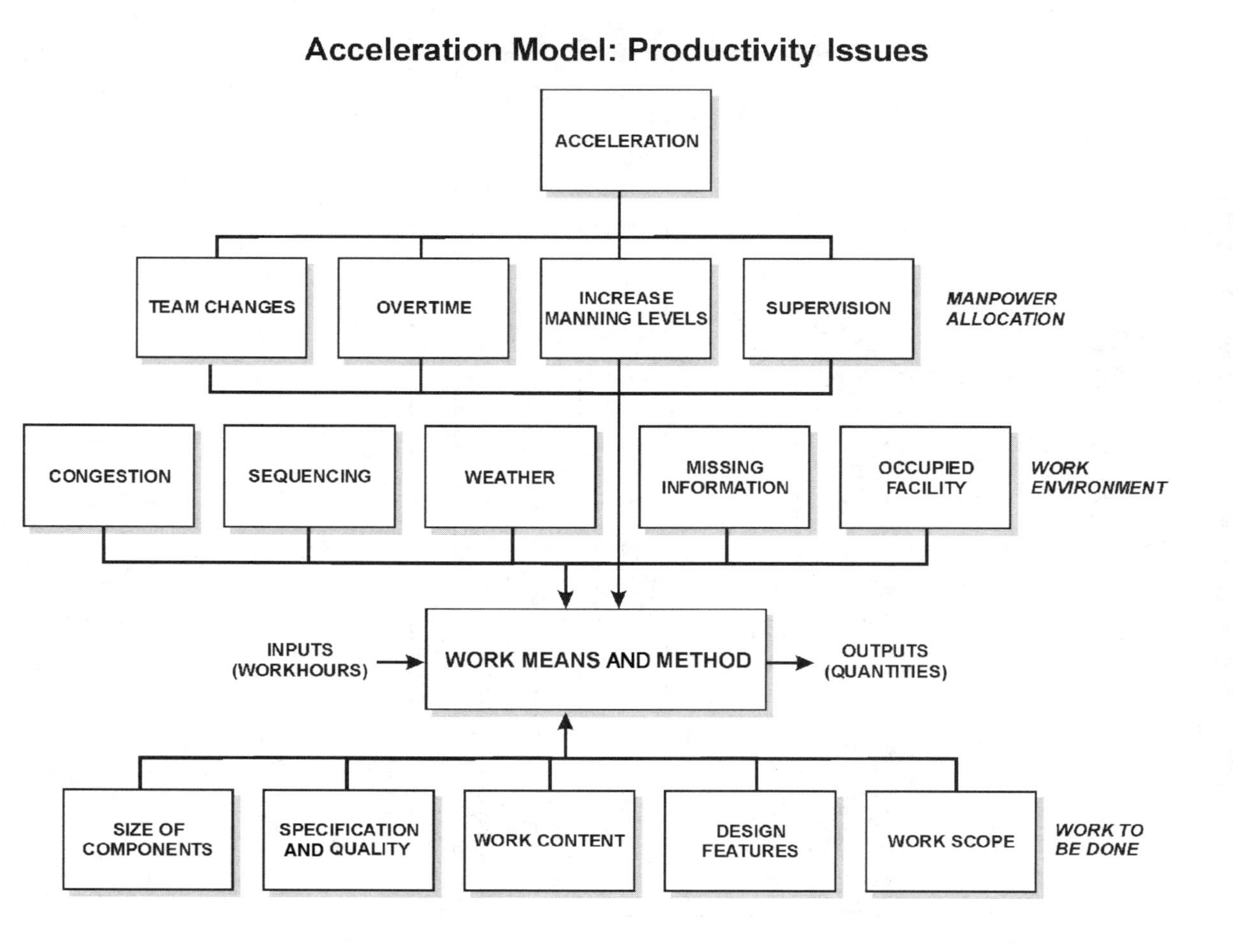

Appendix 17

Measured Mile

Period	Total $ Earned	Work Hours (WH) Used in Period	$/WH
Bid	13,959,000	123,508	113*
February	269,601	1,684	160
March	593,765	3,518	169
April	1,482,949	6,073	244
May	1,396,387	9,834	142
June**	1,751,567	18,022	97
July**	720,987	14,342	50
August/ September**	2,274,534	38,441	59
October	799,291	6,101	131

* Over entire contract duration.

** Project acceleration (overtime) period.

Index

A
At risk 6–8
Accelerate 36
Access 24
Accuracy 31
Ambiguity 76
Analysis 18
Attenuation 16

B
Baseline 15
Beneficial use are 74

C
Cash flow 4
Choices 22
Claims 11
Communication 18
Claim 18
Competing goals 62
Competitive bids 33
Composite goals 10
Concurrent delays 50
Consequence 11
Constructability 42
Construction completion 36
Construction manager 7
Constructive acceleration 77
Contingency 31
Continuous improvement 11, 16
Contractor 6
Control 16
Coordinated 33
Cost plus 23
Cost variance analysis 44
Customer 4
Customer satisfaction 11

D
Delivery method 7, 22
Damages 81
Dependencies 54
Dependent 21
Design/Bid/Build 7, 8
Design/Build 7, 8
Design/build/operate 8
Design professional 5
Development process 29
Directed acceleration 77
Direct risk may 16
Disputes 11

E
Elemental goal 10
Elements 22
Engineer/Procure/Construct (EPC) contractor 8
Entitlement 80
Equal priority 20
Errors and omissions 79

Escalation 32
Exculpatory clauses 75
Excusable delay 50

F
Final completion 74
Fixed cost 23
Fixed price contracts 8
Flexibility 16
Force account 75

G
General contractor 6
Goals of 10, 22
Goals matrix 10
Guaranteed maximum price 8, 23

I
Impacted 82
Independent 21
Indirect risk 16
Indirect risks 15
Inefficiency 82
Integrity 37

L
Liquidated damages 74
Lost opportunity 11
Lump sum 23

M
Management responsibility 23
Manshul Method 81
Measured Mile 82

N
Negative variances 17
Nonexcusable delays 50
Notification 18

O
Operational 36
Oversight 67
Owner 5
Owner's representative 8

P
Prime contractor 7
Performance 79
Performance 80
Performance standards 8
Phases 17
Physical progress 35
Positive variances 17
Pre-commissioning 36
Priorities 20
Prioritization 15, 18
Prioritize 20
Prioritizing risks 54
Priority 21
Probabilistic 19
Probability 11
Productivity 36, 82

Q
Qualitative 18, 25
Qualitative risks 11
Quality assurance 16
Quality control requirements 24
Quantification 18, 25
Quantify risk 19
Quantitative 19

R
Rating 18
Reimbursable 8, 23
Resources 4
Risk management log 17
Responsibility 18

Risk 11
Risk acceptance 19
Risk assessment techniques 28
Risk attenuation 41
Risk management 29
Risk management process 16
Risk sharing 41
Risk transfer 41

S
Scope definition 27
Sequential phases 33
Sharing 16
Specialty contractor 7
Subcontractor 6
Substantial completion 73
Success 16

T
Team decisions 20
Timeliness 17, 23
Total float 20
Tracking progress 35
Transfer 16
Trends 16, 21, 44
Turnkey contractor 8

U
Unimpacted 82
Unit prices 23, 74, 79
Unit price 75
Unit Rate 83

V
Value engineering 42
Variable goals 15
Variance 21
Variances 15

W
What-if analyses 33
What-if analysis 36
Wicks Law 78